Advance Praise for WILD SOULS

"What is wildness? How do we resolve conflicts between the needs of individual animals and the work of preserving species? *Wild Souls* asks readers to think deeply about these and other important questions around our relationship with wildlife. Everybody who cares about animals should read this fascinating book." — Temple Grandin, author of *Animals in Translation* and *Animals Make Us Human*

"In this profound and philosophical book, Emma Marris examines the fiction of a primeval world untouched by human intervention. We have messed with the world in such complex ways that the notion of wildness is at best speculative and at worst entirely artificial: wildness is permitted to exist in designated areas; animals are bred in captivity to repopulate what were once their natural habitats; endangered species are tagged and followed, prioritized over others. In luminous, captivating prose, Marris plumbs the contradictions of our often foolish attachment to the world not as it is, but as we would like to imagine it into being. This is a deeply felt and deeply thought book, brimming with compassion and rue, that throws out revelations like a stream of arrows, each one aimed at the very heart of the matter." — Andrew Solomon, National Book Award-winning author of *The Noonday Demon* and *Far from the Tree*

"Thoughtful, insightful, and wise, *Wild Souls* is a landmark work. With thorough reporting and piercing moral clarity, Emma Marris forces us to think deeply about every aspect of our relationship with wild animals, and what the concept of wildness even means. It should be a guidepost for our thoughts and actions for decades to come." — Ed Yong, author of *I Contain Multitudes*

"In *Wild Souls*, Marris asks the thorny, necessary questions for our time: What exactly is our responsibility to the wild(-ish) animals in the world, and why is it so uncomfortable to figure it out? She challenges us not only to do the 'right' things, but to be our most humane selves in the process. This is the best thinking-and-feeling person's guide to sharing the planet that I know." — Florence Williams, author of *The Nature Fix*

"Like many others, Emma Marris loves wild nature. But unlike most of us, she thinks hard about what words like 'wild' and 'nature' *mean*. As Marris journeys from Northwest wolves to rats in New Zealand, she finds answers that are as fascinating as they are unexpected." — Charles C. Mann, author of *1491* and *The Wizard and the Prophet*

"Through stories that marry adventure and philosophy, Emma Marris works to reconcile the jarring truth that sacrificing individual animals is sometimes the only way to save entire species. Ultimately, *Wild Souls* proposes a new framework for resolving the moral dilemmas that arise as we try to be good stewards of a thoroughly humanized world." — Beth Shapiro, Professor of Ecology and Evolutionary Biology, University of California Santa Cruz, and author of *How to Clone a Mammoth*

"*Wild Souls* challenges us to be better citizens of the planet. How do we think about our relationship to other living things on Earth? With an epic sweep worthy of the subject, Emma Marris links cutting-edge science with deep compassion to provide us tools for approaching the decades ahead." — Neil Shubin, author of *Your Inner Fish* and *Some Assembly Required*

"In this masterpiece of environmental philosophy, Emma Marris cross-examines every claim and subverts every shibboleth of modern conservation. *Wild Souls* brings razor-sharp reasoning and unflinching moral clarity to a field that occasionally suffers from fuzzy logic. This is a book meant to be argued with, in the best possible sense." — Ben Goldfarb, author of *Eager: The Surprising, Secret Life of Beavers and Why They Matter*

"Eloquently, skillfully, Emma Marris wrestles with the dilemmas that define our relationships with animals and the environment, emerging with provocative but necessary answers. I dare any nature lover to read this book and not come away profoundly changed." — Douglas W. Smith, Senior Wildlife Biologist, Yellowstone National Park, and Project Leader for the Yellowstone Gray Wolf Restoration Project

"Where do wild animals fit in a human-dominated world? The answer, for better or worse, will be determined by humans. Emma Marris's exploration of this question is at once thoughtful, thought-provoking, and thoroughly absorbing." — Elizabeth Kolbert, Pulitzer Prize-winning author of *The Sixth Extinction* and *Under a White Sky*

Wild Souls

Rambunctious Garden: Saving Nature in a Post-Wild World

Wild Souls

Freedom and Flourishing in the Non-Human World

Emma Marris

BLOOMSBURY PUBLISHING

NEW YORK · LONDON · OXFORD · NEW DELHI · SYDNEY

BLOOMSBURY PUBLISHING
Bloomsbury Publishing Inc.
1385 Broadway, New York, NY 10018, USA

BLOOMSBURY, BLOOMSBURY PUBLISHING, and the Diana logo are trademarks
of Bloomsbury Publishing Plc

First published in the United States 2021

Portions of chapters 4 and 5 first appeared in somewhat different form in *The Routledge
Companion to Environmental Humanities*. Portions of chapter 10 first appeared in
somewhat different form in *National Geographic*. Portions of chapters 11 and 14 first
appeared in somewhat different form in *Wired* magazine. Portions of chapter 13 are © 2018
Emma Marris, as first published in the *Atlantic*. Portions of chapter 14 first appeared in the
essay "For Whom, the Mammoth?" co-authored with Yasha Rohwer and originally published
by the Center for Humans and Nature as a response to the Questions for a Resilient Future
Series: How Far Should We Go to Bring back Lost Species? To read more responses to this
question, please visit humansandnature.org.

Bloomsbury Publishing Plc does not have any control over, or responsibility for, any
third-party websites referred to or in this book. All internet addresses given in this book
were correct at the time of going to press. The author and publisher regret any
inconvenience caused if addresses have changed or sites have ceased to exist,
but can accept no responsibility for any such changes.

LIBRARY OF CONGRESS CATALOGING-IN-PUBLICATION DATA IS AVAILABLE

ISBN: HB: 978-1-63557-494-4; eBook: 978-1-63557-496-8

2 4 6 8 10 9 7 5 3 1

Typeset by Westchester Publishing Services
Printed and bound in the U.S.A. by Berryville Graphics Inc., Berryville, Virginia

To find out more about our authors and books visit www.bloomsbury.com and sign up
for our newsletters.

Bloomsbury books may be purchased for business or promotional use. For information
on bulk purchases please contact Macmillan Corporate and Premium Sales Department at
specialmarkets@macmillan.com.

For Yasha, my favorite animal

All flourishing is mutual.

—ROBIN WALL KIMMERER

CONTENTS

1

The Flight of the 'Akikiki

For my fortieth birthday, I went to see extinction in paradise.

On a clear, windy December day, I reported to a helicopter landing zone at Lihue Airport on the island of Kaua'i, where several ecologists in flight suits were waiting for a ride. Around us fluttered red-crested cardinals, striking birds with gray backs, white breasts, and cherry-colored faces and crests. These beautiful songbirds are native to South America. They were introduced to Hawai'i in 1928, just one of more than 50 species of birds that have established themselves in the archipelago since humans arrived. In the lowlands, it is these newcomers who are most visible and numerous as they flit among plants introduced from Brazil, Australia, Madagascar, and elsewhere. To see the original inhabitants, we must ascend into the mist-shrouded mountains.

When the Hawaiian islands were born in volcanic convulsions millions of years ago, there was no life on them. They were bare rock. Until the first humans arrived—quite recently, in geologic terms—every species that lived on Hawai'i could be traced to an ancestor that somehow made it there across thousands of miles of ocean. All the birds were descended from about 27 species that flew, or were blown, to these islands in the distant past. Once they arrived, their offspring gradually evolved to eat new foods and thrive in their new habitat. The common raven made it to the islands and evolved into the smaller Hawaiian crow, or ‘alalā. Over time, a single species of honeycreeper gave rise to 47 different species. Back then, Hawai‘i lacked mosquitoes, so the birds stopped maintaining immunological defenses to the many diseases those insects carry.

A red helicopter touched down, just briefly, and three ecologists clutching dry bags scrambled aboard. The pilot never even cut his engines, and they were off again into the sky. A car pulled up and a small curly-haired woman emerged: Lisa "Cali" Crampton, leader of the Kaua‘i Forest Bird Recovery Project since 2010. She handed me a flight suit and zipped one up herself, looking very much like a determined military officer about to go on a special mission. In this case, her mission is saving several species of birds from the abyss of extinction. The helicopter returned and we scuttled over to it, hunched over, and hauled ourselves in. We lifted off and swung into the blue, fields and roads giving way to impossibly steep mountains draped in green foliage and shining with waterfalls. Next stop: Bird Camp.

Humans discovered Hawai‘i during the golden age of Polynesian exploration, between 800 and 1,000 years ago. They brought a whole ecology with them: livestock; dogs; and plants for food, medicine, fiber, wood, and decoration. They also learned how to use local resources, including the beautiful feathers of birds like the bright red ‘apapane.

There were some extinctions when humans first came to Hawai‘i, but over time the people developed complex systems of land and sea management that allowed them to live sustainably on the islands at high densities. For example, bird catchers, known as kia manu, were said to have trapped birds with bait and sticky sap and released them after plucking a few feathers. So it was not creating cloaks for royalty that pushed Kaua‘i's forest birds to the edge of extinction.

No, it was the mosquitoes—and the viruses inside them—that came with a British ship in 1826. "My birds have no immunity to introduced diseases," Crampton said.

The Kaua‘i Forest Bird Recovery Project is trying to save eight different bird species, all hit hard by avian malaria, avian poxvirus, habitat loss, and introduced pigs and rats. Populations of some native birds on Kaua‘i are crashing so fast that their songs are disappearing. There aren't enough older birds to teach the younger birds the melodies.

Two species, the ʻakikiki and ʻakekeʻe, are closest to extinction. The akekeʻe is greenish yellow with a crisscross bill, which it uses to prod and poke buds of the bright red pom-pom flowers of the ʻōhiʻa tree in search of insects. It only lives in Kōkeʻe State Park and the Alakaʻi Wilderness Preserve—where Bird Camp is located—and there are fewer than 1,000 individuals left. The ʻakikiki are tiny puffballs of white and gray, pink-billed and big-eyed. They eat invertebrates they find by pulling loose bits of bark off trees. There are fewer than 500 alive.

These birds only survive high in the mountains, where it is too cold for mosquitoes to reproduce. But as the climate warms, the mosquitoes keep moving higher and higher. Mosquitoes have reached Bird Camp, but at the moment they are only here during the summer. The very slightly colder winter months are enough to kill them off. The birds get a reprieve. But that might not last for long.

We hovered over what looked like an impenetrably thick jungle, and I thought, "There's nowhere to land," until at the last minute I saw a tiny clearing with a very small homemade wooden helicopter "pad." We put down and a second later the helicopter was away. Up this high, it was chilly. The air was filled with a fine mist, blurring the edges of everything. ʻŌhiʻa trees were covered in lichen and moss. We set out along a boardwalk that barely emerged from what seemed to be a bog. Everything was sopping wet.

Crampton led me on a hike through the forest. We waded across a river while holding on to a guide rope, then headed upward on a twisty trail. Our goal was to check rat traps. Mosquitoes may not be up this high year-round—yet—but rats are, and they love to eat songbird eggs and chicks. Crampton said the area we were walking through was surrounded by a pig fence, since pigs consume fruiting shrubs that birds like the puaiohi—a secretive thrush—depend on. The forest feels incredibly remote. Although we are, as the crow flies, around 10 miles from the town of Waimea, the terrain is so steep and trackless that the ecologists stationed here see no one else for weeks at a time, and all supplies come in by helicopter.

Crampton is researching techniques to kill mosquitoes up here. She runs captive breeding efforts to create backup populations for some species. She's also looking at the possibility that the 'akeke'e might have to be moved even higher up on the mountain. And she kills rats.

As we hiked, grabbing onto volcanic rocks and vines for support, I asked Crampton how she keeps doing this work, given the not insignificant possibility that the birds she has devoted a decade to will go extinct. "Logistics," she answered—coordinating volunteers, raising funds, following the latest developments in mosquito control, designing experiments, writing papers. She keeps extremely busy. "We don't have time to get depressed," she said.

We reached the first rat trap. Crampton is using the latest and greatest rat-killing technology, the Goodnature A24, invented in New Zealand. This $200 trap looks unlike any mousetrap you have ever seen. It is essentially a bolt gun—a tiny version of those used in commercial slaughterhouses to kill cattle—inside a baited plastic dome. The gun is powered by a canister of CO_2, similar to those in home seltzer machines. The rodent smells the bait—Crampton uses chocolate-covered coconut—and pokes its head into the dome, triggering the gun, which instantly kills it. The rodents falls away and the gun automatically resets itself. One of these traps can kill 24 animals before the CO_2 canister is changed.

I'd seen these traps in action before, most spectacularly while hiking along a knifelike mountain ridge a few days prior with André F. Raine, project coordinator of the Kaua'i Endangered Seabird Recovery Project. Raine's mission is very similar to Crampton's, except he focuses on sea-going birds such as Hawaiian petrels—called 'ua'u in Hawaiian after their call—and Newell's shearwaters, also known as 'a'o.

En route to get the camera cards out of some burrows he was monitoring, we found a Goodnature trap surrounded by a penumbra of rat corpses in varying states of decay—at least seven by my count. All that was left of one was a gleaming white spine attached to the rat's hairless tail.

"None of it is nice," Raine said, poking at a rat corpse with his boot. "It is a messy, horrible business. But what happens to these birds is also appalling."

His birds don't breed until they are five or six years old. Then they finally choose a mate and dig a burrow high on a cliffside or in a mountain forest, laying eggs that they will sit on patiently for two months. When their babies hatch, they head out to sea to catch squid or fish for them. Often, they return to a nest empty except for blood and feathers.

Goodnature traps come with a clicker that displays how many times the trap has been activated. Given the moist warmth of this tropical forest, as well as the cheerful cannibalism of rats, there isn't always much evidence left if a rat was killed some time ago. Crampton has 425 of these traps deployed throughout the forest—not enough to eradicate the rats, but enough to "knock 'em back" and give the birds a chance. We saw no signs of any rats around the trap, but the counter listed one. Crampton wrote it down in her field notes—the only obituary that rat will ever get.

Farther along, Crampton was in midsentence when she abruptly stopped. She had heard a telltale chirp. We froze. There was some rustling in the olopua, a Hawaiian olive with purple fruits. It was an 'akikiki. It flitted about, stressed out about our presence in its territory. Crampton had her binoculars welded to her face, trying to see the colors of the plastic bands around its leg so she could note the sighting in the research logs. The 'akikiki was moving so quickly she was having trouble. "I think it is white over silver," she said. I jotted this down in my notebook. "It is young," Crampton added. "If there are always young birds on a territory, then the adults are migrating . . . or dying."

In this moment, I knew I was seeing something special. An individual from a very rare species—a species that may not exist a decade from now—was fluttering around us. I searched inside myself for a Big Feeling, but I sensed I was forcing it. The 'akikiki was charming, but as a small grayish songbird it looked and sounded a lot like species I see every day at my bird feeder in Oregon. But then I looked at Crampton watching the bird, her whole body vibrating with excitement, and my heart leapt into my throat.

———

Many people who identify as environmentalists also love animals, and I have always counted myself as a member of both clubs. I absorbed the importance of protecting the environment from the culture I grew up in: a broadly outdoorsy, politically liberal Seattle circle of friends and family. I went to Audubon Day Camp in the summers and learned how to tell a Douglas fir from a western red cedar. I went car camping with my family and even backpacked across the Cascades when I was around ten years-old. I wrote nature poetry. I accepted as given that wilderness was worth protecting, that extinctions were a tragedy, that biodiversity was important.

My personal experience with individual animals was not extensive. My mother was not the pet type, but she once made an exception when our local bookstore was giving away a tortoiseshell kitten. Harriet was perhaps even less of a human-lover than my mother was a cat-lover. Among her favorite pastimes was pressing herself against a riser on our staircase so that she would be invisible to anyone descending. As soon as you put your foot on her step, she would slash at your ankles with her claws. When my parents divorced, one of the conditions my mother insisted on was that my father take the cat. He did, and she lived a long, busy life, only parts of which he was privy to.

Apart from Harriet, my early ideas about animals were mostly formed by books, wildlife documentaries, and the zoo. My grandmother took me and my brothers to Seattle's Woodland Park Zoo often; it was just a mile from our house. From our backyard, we could even hear the hooting of the zoo's siamangs—large, loud gibbons from Southeast Asia. The Woodland Park Zoo was an early leader in presenting animals in naturalistic "habitats" and positioning zoos as champions of nature conservation. Although I couldn't have really explained it to you, as a kid, I felt sure that the zoo was somehow saving wild animals.

The zoo trained me to see animals primarily as instances of their species, and to value them more if they belonged to a rare species. On a class trip "behind the scenes," I even got scratched by a snow leopard. My classmates were jealous, and I experienced a strange thrill knowing that an *endangered species* had touched me. It didn't feel at all the same as being unpleasantly surprised by Harriet's claws while carrying a laundry basket down the stairs.

After graduating from the University of Texas at Austin with a degree in English in 2001, I returned to Seattle and got a job as a secretary at the University of Washington's botany department. During my short stint there, botany merged with zoology to become biology. I remember well the apprehension of the botanists, who were, it must be said, generally quieter, less aggressive, and nicer than the zoologists. They didn't relish the idea of being bedfellows with the strong personalities who studied predators and other charismatic animals. Our office moved across the street to a different building and the smell of the food they gave the fruit flies made me queasy. I realized that I identified with the botanists.

I left that job to study science writing, and in 2005 I started working as a reporter for the journal *Nature* in Washington, DC, with ecology and conservation as part of my beat. I wrote a lot of stories about animals, in part because readers loved them, but I always tried to convince my editors to cover more plant science. I still identified with the botanists.

Most ecosystems comprise animals as well as plants, though, and you can't fully understand one without looking at the other. In my 15-year career covering environmental science, I've been lucky to have close encounters with wombats and wolves, European bison and howler monkeys, humpback whales and Galápagos tortoises, wallabies and takahē. For many of those years, I causally assumed that conservationists, who work to save species, were the best human friends that wild animals had. After all, they try to stop animals from going extinct and they preserve their habitats.

In the early years of my career, I found myself questioning many assumptions about "nature" and "wilderness" that were common in conservation at the time. Were all introduced species really bad? Was there any true "wilderness" left? Did the concept even make sense in a world where Indigenous people shaped ecosystems for thousands of years before European colonists arrived? These questions would lead me to write my first book, *Rambunctious Garden: Saving Nature in a Post-Wild World*. Broadly, I concluded that conservation must focus on protecting the ability of ecosystems to adapt and change in a changing world, rather than attempting to stop or reverse all change.

In more recent years, I've increasingly reported on specific cases where the interests of individual animals seem to conflict with the goal of biodiversity preservation. In order to save species, conservationists kill a surprising number of individual animals. And they treat animals very differently depending on whether they are common or rare; "invasive" or native; domesticated, "feral," or "wild."

―――――――

It was when I moved to Oregon in 2013 that I really began to examine how conservation did and did not make the lives of individual animals better. My husband got a job teaching philosophy at Oregon Tech, a small polytechnic university in Klamath Falls, a former timber town of about 20,000 tucked between the conifer forests of the Cascade Mountains and the vast, arid Great Basin.

Once we moved in, I looked for stories to tell in my new stomping grounds. One story that was in the local paper nearly every week was the return of wild wolves to the area, two generations after they had been intentionally eradicated from most of the United States by a mass poisoning campaign. At the time there were only about 60 wolves in the state. They were (and are) frequently hazed away from livestock and occasionally shot by state officials if they got a taste for sheep or cattle. Some were shot by poachers. Environmentalists wanted them left alone to thrive and multiply, thrilled to have them back in Oregon. "For many, wolves are a symbol of freedom, wilderness, and the American west," the conservation group Oregon Wild wrote.

Just before I moved to Klamath County, so did the most famous wolf in Oregon, a handsome, long-legged gray wolf given the designation OR7 by state biologists and known as "Journey" by local conservation groups. In 2011, OR7 was tracked by state biologists as he walked from the far northeast of the state all the way across the California border, becoming the first wild wolf in the Golden State for almost a century. All told, the young male walked more than 1,000 miles. In 2013, he wandered back over the border to Oregon and the next year he found a mate and had some pups, forming the Rogue Pack, with a territory straddling Jackson and Klamath counties.

OR7 was the talk of the town, a subject of chatter at dinner parties and at the grocery store. A new friend *swore* he had crossed her lakeside property. Most ranchers were not too pleased by his presence, especially when the Rogue Pack started killing young steers in an aspen-studded valley just south of Crater Lake, but many neighbors I talked to were intrigued by the wolves and even respected them, in a way. Friends who worked for federal and state land management agencies in the area were downright gleeful. Suddenly, Klamath County felt a lot wilder.

But how wild was OR7 really? Thanks to his GPS collar, the state of Oregon knew exactly where he was at all times. State biologists had samples of his DNA and could trace his ancestry. He had both a name *and* a nickname. He even had a Twitter account, maintained by some of his fans. If the Oregon Department of Fish and Wildlife wanted to, they could easily have killed him—that is, until his collar went dead and he refused to walk into any of the traps biologists set for him in an effort to replace it.

I began to wonder whether animals that were this heavily monitored and controlled were really *wild*. Was everything that wasn't a pet or livestock "wild"? What about animals that weren't pets or livestock but that were controlled or managed by humans? Reintroducing wolves, in a world on its way to nine billion people before the century's end, means managing the wolf to keep conflicts with humans to a minimum. And it seems hard to argue that management by humans does not decrease wildness. For what is wildness if not the absence of human control?

In Norse mythology, there is a story of a supernatural wolf called Fenrir. He is the god Loki's son and, at first, he lives with the gods. But he becomes worryingly large and powerful and they decide to bind him. After he breaks free from a series of increasingly robust physical chains, the gods control him by binding him with Gleipnir, a magic fetter of paradoxes: the breath of a fish, the beard of a woman, the sound of a cat walking. These dreamlike, impossible ideas seem analogous to me to the intangible laws, rules, and political boundaries that determine where wolves exist today. I've always wondered why the gods didn't just kill the threatening Fenrir, why they preferred a bound wolf to a dead wolf. Today, using GPS collars and tranquilizer darts and "wolf plans" as our fetters, we seem to have made the same choice.

The more I wrote about wolf reintroduction, the more I began to feel uneasy about it. Bringing wolves back may have changed dynamics in some ecosystems, and it certainly made many people happy, but how happy were the wolves themselves? Because we were asking them to live in between human settlements and on the same lands as free-roaming herds of cattle, we were constantly trying to track them and modify their behavior. Some wolves were captured up to five times. And if they didn't follow our rules, we killed them.

According to Mark Hebblewhite, a wolf researcher at the University of Montana, the most common causes of death for wolves in most parts of the United States and Canada are trapping, hunting, poaching, car accidents, and culling by wildlife managers. In a study he did of 22 radio-collared gray wolves that died between 1987 to 2018 in and around to Banff National Park, 90 percent of the deaths were caused by humans. Just three died in "natural" ways: one in an avalanche, one falling from a cliff, and one from causes unknown.

In Oregon, where I live, there are still fewer than 200 wolves total. In 2019, at least seven of them died. One older female died of a bacterial infection, five were hit by cars, and one was legally shot by a rancher because it was chasing his herding dog. I only know one place in the United States where "natural" wolf death is the norm: Yellowstone National Park. Data on 155 deaths of collared wolves between 1998 and 2010 showed 37 percent had been killed by other wolves. Another 27 percent died of "unknown natural causes." If most wolves outside of National Parks die young because of human actions, I think it is legitimate to ask whether having wolves in the West is worth the cost to individual wolves.

Why do we have such different rules for how we treat wild animals versus how we treat our pets and livestock? The question goes beyond wolves. Even in carefully managed parks, wild animals like deer frequently starve to death or are eaten by predators that we lovingly reintroduced in an effort to restore ecosystems to the way they once were. Every day, wild animals die excruciating deaths that would be considered animal cruelty if we let them happen to our horses or dogs.

As a conservationist, I had long been comfortable with the suffering of individual animals in "the wild." But with humans increasingly taking active management roles in "the wild," the premise that we had no ethical obligations to the animals there seemed harder to maintain. If we reintroduced the wolves and managed their numbers and whereabouts, it seemed to me that we were in some way responsible for their welfare and maybe even for the deer they preyed upon. But if that was true, then what about animals whose lives are shaped by us unintentionally by climate change, land development, and species we have moved around? Would they be our responsibility too? The thought induced a kind of intellectual vertigo. Could humans possibly have ethical obligations to all the untold millions of animals on Earth, to every sparrow and ground squirrel and city rat and white-tailed deer? I was overwhelmed.

I began to delve into the vast body of thought and writing about human ethical obligations to other animals, but I found that much of it focused on pets and farm animals. Most of the smaller number of works about our relationship with wild animals tend to assume that they are completely independent of humans, that they live their lives somewhere "out there" beyond the influence of human civilization. Our ethical obligations to wild animals are often presented as being straightforward: we should simply leave them alone and protect their habitat.

The thing is, there is no more "out there." The whole Earth is like a larger version of Kaua'i, with its flora and fauna from all over the planet, legacies of human management going back hundreds of years, and rare animals barely hanging on to existence at the fringes in ecosystems that are warmer and weirder than they once were.

In my previous book, I challenged the idea that there is such a thing as pristine wilderness in the twenty-first century. Humans have dramatically changed the entire world. Starting thousands of years ago, we've changed ecosystems with fire, driven some species extinct, and domesticated dozens of others. In modern times, we've cut down vast forests, converted grasslands to croplands, diverted rivers, and moved mountains. We've built cities, polluted fresh and salt water, sprinkled plastic over everything, lit up the night with artificial lights, filled the air and seas with the noises of billions of machines, crisscrossed the continents with roads, moved plants and animals to new places, and significantly transformed the climate. These changes affect wild animals even hundreds of miles from the closest human settlement.

We've touched many animal species so deeply with our wholesale reshaping of planet Earth that we have likely altered their evolutionary trajectories. I wanted to know whether the massive human impact on Earth changes our obligations to animals. What about animals, like the polar bear, that have lost their hunting grounds because of melting sea ice? Do we have an obligation to feed them? What about wild wolves who mate with feral dogs? Should we stop them? What about introduced mice preying on rare seabirds? Should we poison them? In a human-altered world, it seems impossible to just keep saying that our only ethical responsibility to "wild" animals is to "let nature take its course." It was still unclear to me, though, exactly what this enhanced responsibility might include. Should we be, in some sense, caring for all wild animals? But if we do, will we make them even less wild, less free?

If we could better understand our ethical obligations to our non-human kin, it could significantly improve the way we make decisions in conservation and wildlife management and even in fields like urban planning, veterinary science, pest control, or agriculture. At the moment, whether we legally protect an animal or blithely put it to an agonizing death depends more on the context of the action and the rarity of the species than on whether the animal can feel pain or suffer. Our rules and mores for interacting with animals are capricious and self-contradictory. We can do better.

Some changes must be made in policy or law, but others can be made by individuals. With a better understanding of the ethical choices we are making, we'll be better equipped to decide whether or not to buy an exotic pet, to visit a zoo, to hunt for meat or trophies, or to trap "invasive" species in our backyards.

Non-human animals are different from us, and we can never completely know what it is like to be them. And yet we can love them with a pure, simple love it is sometimes hard to have for other humans. We can be overcome with awe in their presence. They can terrify us—the coiled cat in the night, the howl outside the tent. Our emotions about animals have always been strong, but are our intuitions about how—and whether—to interact with them still correct?

I decided I needed to turn to the experts.

———

Moral philosophy is the study of ethics. Moral philosophers put forth theories about what it means to be a virtuous person; about what is really morally valuable and what is not; about how we, as human beings who want to do the right thing, should act in any given situation. So if I wanted to figure out what I ought to do in regards to wild animals, it was to philosophy that I should turn.

Unfortunately, moral philosophy doesn't have unequivocal answers. Many different ethical theories have been developed, and even within these, you'll find thinkers with divergent arguments. In some ways, philosophy is more about organizing questions than it is about providing practical advice. So while I began my search for answers by *reading* philosophy, I necessarily had to continue it by *doing* philosophy. It turns out, neither a PhD nor a toga is required. In fact, the way I did it involved getting covered in mud checking petrel burrows, sitting around campfires, touring genetics laboratories, and peering inside rat traps. I traveled to the blood red sands of the Australian Outback to find the endangered bettong, to the cocoa-colored river highways of the Peruvian Amazon to hunt for spider monkeys with bows and arrows, and to an uninhabited New Zealand islet to search for a rare and beloved rat.

I began by looking at animals themselves and at the ecosystems they inhabit, trying to really understand the nature of these things I love. Then I looked at ways humans have tried to express their love for wild animals through capture and control—by owning them as pets and displaying them in zoos. I then turned to the conservation work I have long covered as a journalist, looking at captive breeding, supplemental feeding of wild populations, and killing "invasive" species—but this time, I tried to look at these activities through the eyes of the individual animals as well as the framework of protecting species. Along the way, I reassessed meat hunting, one of the oldest relationships between humans and non-humans.

As I take you along on my journey, I hope you'll embark on your own, using these stories and the philosophical approaches I share to investigate your own values and beliefs—and maybe even change them. We will focus on "wild" animals in this book, so I won't be talking too much about dogs or cats or farm animal welfare or going vegetarian, although the lines between pets, livestock, and "wild" animals are blurrier than you might think, as we will see.

I won't be focusing directly on human social justice here either, but it intersects with almost all "conservation" questions. Questioning whose values count and whose do not, looking at who has the money and the power and why, at who benefits from actions taken in the name of "nature"— these are all key questions. I think that ethical treatment of the non-human world likely depends on the ability of historically oppressed peoples, including people of color and Indigenous people, to wield power and bring their values and ways of relating to the world to the conversation on equal terms with the old white guys who currently dominate conservation.

I will tell you upfront that I will not leave you with a mathematical formula for making ethical decisions. There are some cases where no option seems unproblematic, as when, for example, we have to choose between hurting specific animals and losing species. That tension hinges on trying to compare two very different things: the value of individual creatures and the value of complex ecosystems. In some ways, this is the toughest problem of all, and we will turn to it near the end of our journey.

I start with "wild" animals—animals, which I had, for most of my career, seen primarily as units of a species or nodes of an ecosystem. Now I would try to see them as individuals. This would mean thinking of the 'akikiki I saw not just as a token of an endangered species but as a feathered someone with wants, desires, and plans. It means, too, that I would have to try to think of the rats that had died for its sake in the same way.

2

Our Animal Kin

One summer afternoon a bat flew down my chimney—I had left the damper open—and disappeared into our house. My husband was out of town; my children were small, and I knew that rabies shots are recommended for kids exposed to bats and that they cost thousands of dollars. I also felt a rising empathetic panic. The bat was trapped in a complex structure with very few openings to the outside. To the bat, the house must have seemed like a labyrinthine cave, filled with strange textures and shapes for its sonar to try to make sense of.

I began looking for the bat, creeping around like an interloper, terrified of coming around the corner and encountering its panicked flying, which would make *me* panic, and then surely the *kids* would panic . . . I was out of my depth. Dusk was coming on, and the bat seemed to have found a place to hide. I gave up. I brought my kids into my bed to sleep with me and even locked the bedroom door. As they slept, their little animal heads slightly hot and tacky with sweat in the manner of sleeping human children in the summer, I listened for movement and tried to imagine what the bat was doing and thinking.

The next day, while the kids were watching a movie, the bat emerged from a hole in the wall and swooped over their heads. I called our local wildlife rehabilitation organization, Badger Run, and they sent a young woman with leather gloves and a shoe box. She exuded a calm I envied, found the bat, gently picked it up, and popped it in the box. She promised she would take the animal to a quiet spot and place it on a tree branch, since some bats cannot take flight from the ground. I felt relief for myself and for the bat, and a profound sense of an unbridgeable gulf between us. No matter how kindly disposed I was toward this fellow mammal, he or she was so deeply different from me that we simply could not relate to each other in a way that wouldn't freak us both out.

In 1974, a philosopher named Thomas Nagel published a paper in which he wondered what it would be like to be a bat. Actually, his real concern was with what philosophers call the "mind-body problem." It asks whether we are just brains sloshing around in bodies, or whether we are somehow more. Are our thoughts, feelings, intentions—our sense of ourselves—only the outputs of a blob of tissue encased in our skulls?

Nagel's argument was that for all the fancy brain science that was happening in the 1970s, none of it could really explain consciousness. "The fact that an organism has conscious experience *at all* means, basically, that there is something it is like to *be* that organism," he wrote. As an example of how different and specific it would be to be another organism, he selected the bat—a mammal, but one very different from us. "Anyone who has spent some time in an enclosed space with an excited bat knows what it is to encounter a fundamentally *alien* form of life," Nagel adds. Indeed.

Bats fly. Bats live their lives at night. Bats inhabit a world they sense primarily through echolocation, a world made three-dimensional by the texture of sound waves. We might be able, if we are imaginative, to picture (but even that verb shows our visual bias!) what it would be like to fly at night or even to hear the landscape—we can imagine what it would be like for *us* to be bats. But, Nagel says, "That is not the question. I want to know what it is like for a *bat* to be a bat."

And this, Nagel argues, is simply not possible. His point is that even if we map every neuron in the bat brain, we can still never really know what it is like for a bat to be a bat. And that means that there is something—what he calls the "subjective character of experience"—that cannot be explained by the physical facts of a bat's body.

The seemingly unbridgeable gulf between the human animal and non-human animals has sometimes led humans to believe that since they cannot know what it is like inside bats' heads, there is nothing inside their heads. On the contrary, the very existence of the "subjective character of experience" seems to suggest to me that they have and are *selves*, even if their selfhood is very different from our own.

———

We can never know precisely what it is like to inhabit the consciousness of an animal of another species, but we are learning more and more about them all the time. As every year passes, more scientific studies are published showing that non-human animals are smart, emotional, and even kind. Their inner lives are rich. People who live with non-humans knew this. But science is finally catching up. Each new finding makes it more difficult to argue that animals don't deserve some sort of moral status. If they can suffer, if they can remember, if they can love, if they can choose, then surely we cannot justify treating them like mere things.

The last common ancestor of all the great apes—including humans—lived some 13 million years ago. The rodent-esque common ancestor of all the placental mammals—including humans—lived around 66 million years ago, just after the dinosaurs got nailed by the asteroid. And the wormlike common ancestor of all bilaterally symmetrical animals—including humans—lived some 550 million years ago. We are all related.

Given that we are related to all other animals, the most conservative hypothesis would be that our mental differences from them would be in degree, not in kind. There's simply no reason why there would be a sharp line in either neuroanatomy or cognitive abilities separating humans and non-humans. And in fact, non-humans have turned out to possess nearly every ability that scientists have put forth as being exclusive to humans.

No one seriously doubts anymore that our fellow mammals are sentient. In 2012, a meeting of neuroscientists drafted and signed on to the

Cambridge Declaration on Consciousness, which asserted that "the neuro-logical substrates that generate consciousness" were possessed by "all mammals and birds, and many other creatures, including octopuses." The question of whether *insects* are conscious was reopened in 2016 with the publication of a paper by Andrew B. Barron—a biologist—and Colin Klein—a philosopher. The duo came together to argue that the presence of structures in the insect brain that created a model of itself in space meant that insects had a sense of themselves *as themselves*, and thus a "subjective experience."

Human brains are not the largest in the animal kingdom, nor do they possess more neurons than other species. Experts once claimed that only humans were "self-aware," in the sense of being able to conceptualize themselves as a distinct entity. But that turns out not to be true. To see if animals are self-aware, researchers often surreptitiously put a strange mark on an animal's forehead, then provide it with a mirror. The idea is that if it looks in the mirror and can understand it is looking at itself, it will touch its own face to try to remove the mark. Some elephants do this. Dolphins will do it as well, as will magpies. Chimps do the same, although, as animal behavior researcher Frans de Waal points out, the formal test is hardly necessary to establish that they "get" that they are the ape in the mirror. De Waal has seen chimps use mirrors to check out their teeth, to admire their own butts, examine an injury on the top of their heads, and even to assist them as they clean out their ears with a straw. In 2019, a fish in the wrasse family—a group of clever and adaptable fish with pouting lips—passed the test. After trying various movements in front of the mirror, like Groucho Marx in *Duck Soup*, the captive wrasses evidently understood they were looking at themselves and began trying to scrape the marks off.

Dolphins don't just know themselves as individuals: they have names. Each has a unique call that it repeats at intervals, as if to say, "Joe here!" Dolphins in groups use each other's names to get one another's attention. Dolphins kept alone in captivity call out their own names and never hear a response. Parrots have names too. Researchers think they are given these names by their parents while they are still in the nest. Bats also produce individually unique calls, which they recognize as their own. They learn these calls from their mothers. In a very real sense, the bat in my house had a name.

Many animals have "personalities," with some individuals measurably bolder than others, or more curious than others, or more social than others, or better at solving physical problems. These traits show up again and again in the same individual, even in different contexts. Personality has been demonstrated in spiders, cabbage white butterflies, crickets, and bees. Frogs, toads, salamanders, and newts all show differences from individual to individual in terms of boldness, tendency to explore, and activity level. The fact that animals vary from individual to individual isn't lost on the animals themselves. That's part of why animals have friends—individuals that they prefer to spend time with. In one experiment, researchers watched 10 juvenile eastern garter snakes socialize in an enclosure with four hidey-holes. In the video, the snakes crisscross their enclosure, visiting with one another. The researchers found that not only are the snakes quite social, they also consistently sought out the same snakes, even after researchers shuffled their locations and cleaned the enclosure to remove odor cues. The results suggest that the snakes see each other as individuals and have preferences for some individuals over others. "They have sophisticated social cognition," one of the researchers told *National Geographic*. "They can tell others apart."

Many humans know from experience that mammals like dogs, horses, and cats have distinct personalities. Sometimes, we can even trace how their personality might have been influenced by their life experiences. Researchers have identified core traits that vary between dogs, including playfulness, chase-proneness, curiosity/fearlessness, sociability, and aggressiveness. But undomesticated animals have personalities too. Rats can be introverts or extroverts, bold or shy, aggressive or placid. Some of personality is genetic, which explains how pet-rat breeders are able to select for calm, cuddly rats.

Some say that only human beings have an open-ended language that can be used to discuss the past and future as well as the present, but other animals communicate with each other in relatively complex systems using sounds, gestures, and even wiggles. Honeybees dance complicated directions to hive mates. Birds and mammals warn each other about threats with alarm calls, which can be different depending on the type of predator, the location of the threat, and the urgency of the situation. Apes communicate a range of things with hand gestures, including the universal outstretched hand request, meaning "share." We make this gesture too. And human researchers have taught some animals totally new language systems. Chimps and gorillas can be taught hundreds of words in sign language and will spontaneously combine signs in inventive ways. A chimp called Washoe named a swan with the signs for "water" and bird." A captive African gray parrot named Alex learned to recognize and say over 100 words for objects, using his skills to demand specific lunches, ask questions, point out similarities and differences between objects, count, do simple addition, and to say, "Wanna go tree," to ask to see the tree outside the window of the lab, a view of which he was quite fond.

Many animals also possess some version of a "theory of mind," which means they not only know that they are individuals, they understand that others also have a subjective mental experience like their own. Knowing this, they can imagine what other individuals are thinking. Scrub jays, blue corvids that live in my part of Oregon, are smart enough to realize that other jays may steal their food if they are observed while hiding it. So they will return later, when the second jay isn't around, and move their stash.

A theory of mind doesn't always lead to dissembling. It is also crucial for cooperation and many types of helping. Knowing that their friends and family feel hunger and thirst, chimpanzees will bring food and water to the old and infirm, sometimes carrying the water in their own mouths. Capuchin monkeys will share food with the monkey in the next enclosure—but not if they've just watched their neighbor have a big meal. Rhesus macaques subjected to a fiendish experiment in the 1960s where pulling a chain would produce a food pellet but also shock the macaque in the cage next door stopped pulling the chain. One starved itself for 12 days rather than hurt its fellow prisoner. Dolphins will lift an injured or stunned friend on their backs, making sure he or she can reach the surface to breathe. Orcas were observed doing something similar to a friend who had been hit by a ferry boat. Chimpanzees will hand over a tool to allow a friend to access a treat. They must pick the right tool for the job and have the kindness to pass it along, and they do both easily. In a famous case, a bonobo—a close relative of chimps—named Kuni found an injured starling. She "climbed to the highest point on the highest tree, and carefully unfolded the bird's wings, one wing in each hand, before throwing it in the air."

Rats too will help a fellow in need—say one trapped in a little plastic tube by researchers. In one study, rats—who strongly dislike swimming— will paddle across a pool of water to reach and rescue another rat trapped in a tube. In another test, when rats were presented with a fellow rat trapped in a tube and five chocolate chips, they faced a choice between keeping the chocolate for themselves or freeing the trapped rat and sharing the chocolate. Just over half of the time, they chose to share. I asked researcher Peggy Mason at the University of Chicago, who has devised several of these experiments, whether she was ready to say that rats had empathy. She wasn't 100 percent ready, she said, but she was leaning that way. "The most important piece of evidence is that the helper rat has to feel the full range of emotion in order to be motivated to help," she said. "If we give that rat an anti-anxiety drug, then the rat doesn't see the problem. That really suggests to us that the rat is acting from an emotional motivation." Another clue is the rats' behavior after the trapped rat is released from the tube. The helper rat "follows the liberated rat," Mason says. "He jumps on him and he licks him." When primatologist Frans de Waal came to visit her lab, he pointed out that this looked a lot like "consolation behavior" in prairie voles. "The helper rat realizes that the trapped rat was in distress and is consoling it."

Communication and theory of mind come together in activities like cooperative hunting, seen in primates, whales, canines, felines, and hawks. I was once lucky enough to watch a pod of humpback whales catch fish together using the "bubble net" technique. Between 6 and 12 whales would all dive together, forming a circle deep underwater. They'd then release a stream of bubbles from their blowholes while circling a school of fish. Fish won't cross the screen of bubbles and can be herded into this "net" in the water column. One whale would give a signal, and they'd all open their mouths wide and swim upward through the ball of fish, swallowing them whole. From the surface, we could see the circular agitation of the water made by the bubbles. Seabirds wheeled above, waiting. Then we'd see the birds start to dive as they saw the fish rising toward them. A moment later, whale jaws would burst through the skin of the sea.

Some animals cooperate across species lines. Ravens often tag along with wolves, relying on the powerful canines to kill and open up carcasses that they can then feed on. Some say that ravens even lead wolves to dead animals and hunting opportunities. "Personally, I think ravens do point out moose for wolves, but it's pretty difficult to prove something like that," wolf researcher Rolf Peterson told me in an email. "Ravens are usually following wolves when they are hunting, perching in trees when wolves rest, and continuing on when the wolves get going. Wolves can certainly find dead moose by following the calls of ravens (I have even done that), and it's only a short leap for ravens to call wolves' attention to a live moose that might be in trouble (I expect ravens are quite practiced at figuring that out)."

A camera trap video that went viral in early 2020 shows another cross-species hunting team. In it, an obviously stoked coyote awaits a badger near the opening to a culvert in the San Francisco area. The coyote greets the badger with a classic canine "play bow," putting his head down low and his butt up high, signaling his lack of hostile intent. Then the two of them head down the culvert, likely on their way to do some collaborative hunting. Typically, the badger will dig after a burrowing creature such as a ground squirrel, flushing it up to the surface, where the coyote is waiting to pounce. Squirrels that think the better of emerging into the jaws of a coyote retreat into the path of the badger. Both predators make out well.

There is even an aquatic version of this partnership, in which a moray eel is the badger and a leopard coral grouper is the coyote. The eel wriggles through tight spaces in the reef, flushing out prey, which the trout then can nab. Prey that attempts to evade the trout by seeking the security of the reef end up as eel food.

For a long time, experts claimed that only humans made or used tools, but there's evidence of tool use all over the animal kingdom. Chimpanzees strip the branches off sticks and use them to fish for termites, chew leaves into pulp to act as a sponge, and fashion wooden spears for hunting. Capuchin monkeys use rocks as hammers and anvils to open nuts. Macaques and otters are also wielders of stone hammers. Polar bears throw ice and rocks at seals while hunting, with sometimes fatal effect. Elephants drop logs onto electric fences to break the current. New Caledonian crows are master toolmakers, hunting for grubs in rotten wood with spears and hooks they fashion themselves. Many species of kites and falcons from around the world are known to pick up burning sticks in their beaks or talons and fly them to new areas to spread fire so that they can hunt animals fleeing the flames.

And many animals can invent new tools for new challenges. Elephants will push a block in from another room to use as a stool to get closer to a tasty treat that's beyond their trunk's reach. Chimpanzees will stack up many boxes and use a stick to get bananas near the ceiling. Even pigeons will use boxes as stools. Corvids will toss stones into beakers filled with water to raise the level of the water until they can reach a floating treat.

Each chimpanzee community in the wild uses up to 25 different tools—but each group has a slightly different tool set, adapted to their particular area and passed down through the generations. As science has increasingly seen animals as capable of learning, they've begun to record animal culture—ways of doing things that are only observed in some populations of a certain species. The most famous example of animal culture is among the earliest: the observation in 1953 of an innovative behavior—washing sweet potatoes before eating them—that was invented by a young female macaque named Imo and subsequently spread to the rest of the macaques on the Japanese island where she lived. (It is interesting to note that the researchers studying these monkeys chose to give them names rather than numbers, as was standard at the time; one wonders whether this contributed to the scientists' ability to see this as a cultural trend started by a single individual.)

Orca whales eat all sorts of things—salmon, otters, baby baleen whales, even birds. But distinct "ecotypes" specialize in specific prey. Type B1 orcas in Antarctica mostly eat Weddell seals, while smaller type B2 orcas there eat penguins. There's an even smaller Antarctic type—type C, which has a "forward-slanted eye patch"—that hangs out in the pack ice and eats fish. In the Pacific Ocean there are so-called transient orcas, who range widely. They hunt in small, silent packs and mostly eat whales, seals, and other mammals. These transients studiously avoid the chatty "resident" orcas of the Salish Sea in, who are more sedentary and eat fish. Researchers believe these differences are cultural rather than innate in part because they've observed parents teaching their children how to hunt specific prey. When a whale raised in captivity was released in Icelandic waters, it needed to be fed frozen herring to survive.

Animals also arguably make art. The male bowerbirds of New Guinea and Australia dedicate huge fractions of their time and energy to creating elaborate structures from twigs, flowers, berries, beetle wings, and even colorful trash. These are the backdrops to their complex mating dances, which include acrobatic moves and even imitations of other species. What's most amazing about the towers and "bowers" they construct is that they aren't stereotyped like a beehive or hummingbird nest. Each one is different. Artistic skill, along with fine craftsbirdship, is rewarded by the females. Many researchers suggest these displays are used by the females to gauge the cognitive abilities of her potential mates, but Darwin thought that she was actually attracted to their *beauty*. In other words, the bowers aren't simply signals of mate quality; they are appreciated by the females for their own sake, much as we appreciate a painting or a bouquet of spring flowers. A 2013 study looked at whether bowerbirds that did better on cognitive tests were more successful at attracting mates. They were not, suggesting whatever the females are looking for, it isn't a straightforward proxy for cognitive ability.

Animals definitely feel emotions. Psychobiologist Jaak Panksepp told journalist Virginia Morell that emotions are "evolutionary skills" that help a wide range of species navigate the complex problems in everyday life.

Panksepp says the brain areas associated with emotions are evolutionarily ancient and present in nearly all mammals. The idea that emotions are "skills" was somewhat revelatory to me. As a woman who cries easily and feels emotions strongly, I had long seen my feelings as maladaptive and a little bit embarrassing—as impediments to rational thinking. But emotions are so common in animals that they are almost certainly adaptive.

Joy is old. Our closest kin, the apes, laugh when they are tickled and when at play. Wolves play and frolic when young, though they become more serious as adults. Dogs are Peter Pan canines, never growing up, playing for their whole lives. Baby rats play and laugh and like to be tickled. Rat laughter is a chirp so high-pitched that humans can't hear it. Fear and sadness are widespread as well, alas, nestled into the brain right next to the systems for physical pain. Bird and mammal babies commonly make little distress calls if left alone by their parents too long. The brain regions that provoke these cries are so close to pain centers that Panksepp could silence baby animals by giving them painkilling opiates. There's no reason to believe that they do not feel psychological pain when they make distress calls, just as we do when we are lonely and scared, Panksepp says.

Animals fight and make up, go to war, and mourn their dead. Grief is felt in many species. Elephants use their trunks to caress the skulls and jawbones of lost loved ones, lingering over them for hours, sometimes carrying them for short distances. Chimpanzee mothers who lose their children will carry their corpses for days—in one case for 68 days. "She exhibited extensive care of the body, grooming it regularly, sharing her day- and night-nests with it, and showing distress whenever they became separated," researchers wrote of one such mother.

In one 2018 case, the orca J35, also known as Tahlequah, managed to carry her own dead infant for 17 days. The grieving mother swam with the corpse on her back and the front of her head and even carried the body in her mouth. Her pod mates may have brought her food during her period of mourning.

Many ethicists focus less on an animal's intelligence or rationality than on its ability to feel pleasure and, in the fullest sense of the word, suffering.

Animal welfare researchers draw a distinction between systems that merely precipitate a reaction when a dangerous stimulus is detected and systems that take information about those stimuli into the organism's conscious mind to be experienced as unpleasant. Only the latter is considered real pain, and anyone who has had a pet knows that animals can feel it. When a dog is stung by a bee or a cat catches its paw in a door, its reaction is immediate and unambiguous. Welfare researchers have learned the distinctive signs of pain in a variety of species. Young animals tend to cry out, presumably calling for their parents' protection, but older animals often struggle in relative silence to escape the pain. Lab mice "grimace," squeezing their eyes shut and flattening their ears.

In 2016, my mind was thoroughly blown by Jonathan Balcombe's book *What a Fish Knows*, which paints a portrait of "the inner lives of our underwater cousins." It turns out that fish are complex thinkers with excellent memories. They make plans and can be curious. Fish stroke and caress one another and apparently enjoy being touched just as we do. And they possess the same kind of nerve fibers we have—nerve fibers that transmit pain, which we are almost certain they can feel. In an experiment that haunts my dreams, zebra fish were kept in a tank that had lots of plants and objects to explore at one end and absolutely nothing at the other end. If the fish were just robots, we might assume they would spend the same amount of time on each end of the tank, swimming in a random pattern. In fact, they clearly preferred the "enriched" end of the tank. This preference didn't change when they were injected with acetic acid, the compound that gives vinegar its bite, which in us, would sting and cause pain. But when they were injected with acetic acid and morphine was dripped into the barren, empty end of the tank, they were suddenly much more likely to be found there. "Thus," Balcombe wrote, "zebra fishes will pay a cost in return for gaining some relief from their pain."

Why then, for so long, have humans just assumed—or perhaps hoped—that fish feel no pain and are essentially mindless? Balcombe thinks the problem is our inability to read their expressions or emotions. There's no sympathy trigger. "We hear no screams and see no tears when their mouths are impaled and their bodies pulled from the water," he writes. "Their unblinking eyes—constantly bathed in water and thus in no need of lids—amplify the illusion that they feel nothing." Many do in fact vocalize when they are in pain, but the sound is designed to be heard underwater, and we can't hear it.

There are many reasons that science has taken so long to see the intelligence and feelings behind animals' eyes. There is no doubt that our desire to be able to keep eating them, capturing them, displaying them, and experimenting on them are among those reasons. Another is our own species' limitations. Frans de Waal titled his 2016 book on animal cognition *Are We Smart Enough to Know How Smart Animals Are?* to call attention to the ways in which our human brains have shaped our expectations of animal intelligence. Humans are good at language, cooperation, and toolmaking, so we tend to value these capabilities and test for them. But other species far outperform us in mental abilities more adapted to their own ecological roles. "It seems highly unfair to ask if a squirrel can count to ten if counting is not really what a squirrel's life is about," de Waal writes. Squirrels may fail at the counting task, but they beat the pants off us when it comes to remembering the hiding places of thousands of hidden nuts.

Far from being creatures of pure instinct, carrying out a limited number of behavioral routines like a non-player character in a video game, individual animals live in a massively complex interspecific social world, constantly observing the goings-on around them, making choices, solving problems, finding food, raising young—and even pursuing joy. When you see a bat briefly illuminated in the moonlight, it isn't just "a bat," it is a particular bat, with particular personality traits, memories of its life, a family it cares about, and a plan for at least the immediate future. We cannot know what it is like to be a bat, but we can be sure that bats know what it is like to be a bat. There was an individual swooping around my house that day, a someone.

3

Philosophies of the Non-Human

W e have always been one animal among many. In many cultures, that's still how we see ourselves. In the culture I grew up in, though, I absorbed a different understanding in which humans did not count as animals at all and in which animals were simple creatures driven by instinct. Any obligations to treat them well were obligations of charity and kindness to lesser beings. Both of these narratives—animals as kin and animals as lesser beings without agency—have been with us for a long time.

The narrative of animals as kin is almost certainly older. For most of our evolutionary history, other animals have been an existential threat to us as well as a source of sustenance. In our earliest days, we were likely prey more often than we were predator. Today, primates from tiny mouse lemurs to giant gorillas are preyed on by everything from owls to leopards and tigers. And fossil remains of our ancestors and their cousins show that we were likely no different. A skull from a toddler *Australopithecus africanus*, 2.8 million years old, shows puncture marks just like those made on monkey skulls today by eagles, and the tiny cranium was found in a heap of bones typical of raptor prey. A skull of one of our relatives called *Paranthropus* from about 2 million years ago has been found with two little holes in it—just the size and spacing one might expect from a contemporaneous leopard species. Fossil hominins some 1.8 million years old, found in what is now the Eastern European country of Georgia, display gnaw marks from big cats or hyenas.

Being prey shaped us. Our social nature may derive, in part, from the safety of hanging out in large groups back when leopards were a greater threat to us than viruses. Our many languages may have begun as alarm calls alerting each other to approaching predators. Some researchers think we began to walk upright because it allowed us to look for predators from farther away in grasslands, like big primate meerkats or prairie dogs. Standing up also made us look bigger, which might have helped scare away some would-be attackers. And walking on two feet means it is relatively easy to pick up our babies and run when trouble arrives. These details are still speculative, but it would be remarkable if millions of years of being hunted by other animals *didn't* alter our evolutionary course.

We also ate other animals. At least 2.6 million years ago, our ancestors began eating meat—researchers think we may have first scavenged from the kills of feline predators. We used stone tools to get meat off bones; scientists can still see the marks. Then we cracked the bones and ate the marrow. By 1.5 million years ago, we were hunting animals for meat ourselves.

Given the dual role of non-human animals in our lives as both bringers of death and bringers of life, it makes sense that many human religions have involved worshiping animals or sacrificing animals to gods to gain their favor.

In what is now Indonesia, in what is perhaps the oldest figurative painting in the world, some 44,000 years old, animal-human hybrid figures hunt wild pigs and buffalo. In Chauvet cave in what is now Southern France, paintings from about 30,000 years ago depict mammoths, cave bears, cave lions, wooly rhinos, and aurochs, the ancestors of modern cattle. The heads of the lions overlap, seeming almost to suggest motion in the flickering candlelight of the darkened cave. We may never know what exactly these paintings meant to those who drew them.

Our gods are often part human, part non-human. One of the oldest statues in the world, known as the Löwenmensch figurine, is 35,000 to

40,000 years old, and depicts a human with a feline head. Then there's the mysterious antlered Celtic god Cernunnos, of which we have only images and no stories. He often holds a snake with a ram's head and sits cross-legged on the ground and might have been a god of the "in between," "both wild and tame," according to one expert. The Maya have an antlered god too, Siip (or Sip), an old man with a deer's antlers and ears, sometimes with a wasp nest between his antlers. Siip can transform from one game animal to another, from a deer to an armadillo to a peccary.

Greece had its satyrs, minotaurs, and centaurs. Egypt's gods often had animal heads on human bodies: Horus, "lord of the sky," with a falcon's head, the fierce goddess Bastet with a feline head, the death god Anubis with a jackal head, and the dangerous god Seth with the head of an uncomfortably unidentifiable beast, a creature Egyptologists simply call "the Seth animal." In Hinduism, Hanuman is a heroic figure with a monkey's head and tail, while Ganesha, god of beginnings and patron of scholars, has an elephant's head with one broken tusk.

In some cases, living non-human animals are thought to have sacred properties. Native American peoples on the East Coast of North America at the time of colonization saw some animals, especially those they hunted, as possessors of spiritual power. The brown bears of Hokkaido are both living animals and embodiments of gods to the indigenous Ainu of Japan. Their worship traditionally included the annual ritual sacrifice of a bear. Some Christian sects are well-known for handling venomous snakes as a part of their worship, seeing them as incarnations of Satan. In these religious traditions we see a world where both humans and non-humans—as well as hybrids—could be gods and have great power.

The more familiar Western narrative of humans as the masters of animals goes back at least to the ancient Greeks. Ancient Greek civilization was wildly divided on the question of whether humans had moral obligations to non-human animals. The Stoics, who were to become influential in Christian circles, tended to use animals as metaphors for the "lower" parts of human nature—irrationality, gluttony, lust, and so on. They argued that animals were not only not sentient, they claimed they literally didn't feel anything.

Their astonishing argument was that since non-human animals can't talk, and since only those who can talk have reason, and only those who have reason can really perceive or feel, animals were alive and animated but without minds or feelings. This opinion was probably not widely shared outside of rarefied philosophical circles. Anyone who has ever actually seen a dog hurt knows very well that they can feel. There was also a more mystical counterculture in the ancient Mediterranean, which considered the possibility that the human soul could have flowed through non-human incarnations. Some believers thus renounced both animal sacrifice and eating meat.

As with the Greeks, Christian thought on animals was not and is not monolithic. On the one hand, Christians were explicitly given a role apart from and above animals by God. As it says in Genesis 1:26, "And God said, Let us make man in our image, after our likeness: and let them have dominion over the fish of the sea, and over the fowl of the air, and over the cattle, and over all the earth, and over every creeping thing that creepeth upon the earth."

On the other hand, one finds Christian figures like Saint Francis of Assisi, a medieval cleric famous for his concern for animals as well as the non-human world more broadly. Saint Francis's famous Canticle of the Sun addresses "Brother Sun" and "Sister Moon," trees, flowers, and other growing things. He preached to the birds and negotiated a deal between a wolf and the city of Gubbio, as recounted in the fourteenth century *Fioretti di San Francesco*. If the wolf agreed to stop killing people and other animals, the townsfolk would promise to feed him, "as it is hunger which has made thee do so much evil." To indicate he agreed to the deal, the wolf "lifted up his paw and placed it familiarly in the hand of St. Francis, giving him thereby the only pledge which was in his power." The compact held for two years. The wolf "went familiarly from door to door without harming anyone, and all the people received him courteously, feeding him with great pleasure, and no dog barked at him as he went about. At last, after two years, he died of old age, and the people of Gubbio mourned his loss greatly; for when they saw him going about so gently amongst them all, he reminded them of the virtue and sanctity of St Francis."

Again, we see two competing narratives, one in which humans are above animals and one in which humans are just one kind of animal sharing space—in which non-human animals are seen as agents with whom you can negotiate.

In many cultures, humans can transform into animals, and animals can transform into people—and stories of these transformations contain echoes of a worldview in which the borders between humans and non-humans have been seen as porous. In Japanese and northeast Chinese folklore, foxes transform into beautiful women and seduce and marry men. In Scotland, you are more likely to be seduced by a Selkie—a seal transformed into a woman. If you want to keep her around, you must part her from her seal skin or she will slip it back on and disappear back into the cold, oily sea. In Inuvialuit stories, polar bears can become humans. As one elder tells it, "They put their snout out onto the ground, and their skins peeled off of them." Just as in the story of the Selkie, destroying the bear skin prevents the person from resuming bear form. The existence of werewolves was an accepted fact in seventeenth century Europe. In an Indigenous community in the Peruvian Amazon, I met the widow of a man who was said to have transformed into a jaguar.

People can be animals sometimes, but animals can also be people. The non-human animals in many traditional Native American tales are much like us, talking beings with houses and spouses who tell jokes and go hunting. These animal people suggest a world where instead of humans standing apart from dumb automatons, humans and other animals interact in a whirl of agendas, interests, emotions, relationships, laughter, and tears. It's a compelling vision.

Evidence that this attitude was common even among some Christian communities comes from the fascinating history of non-human animals put on trial for murder, destruction of property, bestiality, and other crimes in Europe and European colonies from the ninth through the nineteenth centuries. Typically, domestic animals, as members of a human household, were tried for violent crimes by secular courts, while wild animals were tried by ecclesiastical courts—enjoined to refrain or desist from infesting churches or devouring crops. In 1487, the Bishop of Autun ordered processions throughout his diocese for three days to warn the slugs living there that if they did not quit eating herbs and grapevines and leave the area, they would be officially cursed. During the sixteenth century, the people of Beaune, France, asked the ecclesiastical tribunal of Autun to excommunicate some insects raiding their crops. The insects were duly cursed with "anathema and perpetual malediction." When insects were put on trial and sentenced to death, a few of their kind would sometimes be ritualistically executed.

Insects didn't always end up on the business end of anathema. In one case in 1545, the wine growers of the town of St. Julien complained that weevils were eating their vines. The weevils were assigned legal representation, who apparently did their jobs well, for the judge ruled that "inasmuch as God, the supreme author of all that exists, hath ordained that the earth should bring forth fruits and herbs, not solely for the sustenance of rational human beings, but likewise for the preservation and support of insects, which fly about on the surface of the soil, therefore it would be unbecoming to proceed with rashness and precipitance against the animals now actually accused and indicted; on the contrary, it would be more fitting for us to have recourse to the mercy of heaven and to implore pardon for our sins." Public prayers and masses were ordered and the altar bread was paraded around the vineyards. That seemed to do the trick.

In 1587 the weevils returned and were prosecuted again. Their lawyer, one Pierre Rembaud, argued that eating plants was "a legitimate right conferred upon them at the time of their creation," citing Genesis, in which God says "to every beast of the earth, and to every fowl of the air, and to every thing that creepeth upon the earth, wherein there is life, I have given every green herb for meat." He also argued that expecting "brute beasts" to follow human law was absurd. Prosecutors countered that Genesis also said that man had dominion over animals. The trial was taking weeks, so the people of St. Julien suggested a compromise, offering to allow the weevils free rein to eat whatever they wanted in a designated area away from the grapevines. The weevil's procurator, or agent, Antoine Filliol, said the preserve offered by the townsfolk was not good enough for his clients. The judge appointed some independent parties to examine the proposed weevil sanctuary and report back on its suitability for the insects, and paid out three florins to cover their travel expenses.

Alas for students of history, the final outcome of the trial is not known, as the last page of the record was partially consumed by animals. But this trial and the arguments presented in it were by no means unique. Even the idea of recommending or setting aside a sanctuary to which the pests—be they rodents or insects—could retire without further molestation was apparently quite common. In 1713, a colony of termites devouring a Brazilian monastery were said to have removed themselves in an orderly fashion to their appointed "reservation" as soon as the judgment was read to them.

Pigs can be dangerous, especially to incapacitated adults, children, and infants, all of which they have been known to kill and eat. And in earlier eras, they were often allowed to run around loose. Thus history records that many pigs were tried for murder after attacking and killing humans. The pigs were held in jail with human prisoners and fed the same rations. After being found guilty, they were typically sentenced to be burned or hanged "until death ensueth." The owners of these criminal pigs faced no consequences and were sometimes even compensated from the public purse for the loss of their homicidal livestock. Cattle, oxen, dogs, and horses were also tried and executed for murder.

Animals were also frequently executed for the crime of bestiality, which seems wholly unfair, as it assumes they were consenting parties to the deed. In at least one case—the prosecution in 1750 of one Jacques Ferron and a female donkey—the jenny was acquitted on the grounds that she "was the victim of violence and had not participated in her master's crime of her own free-will." The inhabitants of the town of Vanvres signed a certificate attesting to the donkey's good character, which read, in part, that "she is in word and deed and in all her habits of life a most honest creature."

All together, these trials suggest that medieval and early modern Europeans ascribed to these non-humans some kind of rationality and some kind of agency. They paint them as *persons* who can do wrong.

During the Enlightenment, though, an elite focus on the power of logic and rationality led to the creation of an increasingly bright conceptual line between humans, who reasoned, and animals, who were thought to be devoid of any independent thought.

French philosopher René Descartes believed that the physical world, including animal bodies and brains, was made of matter and followed mathematical laws. The exception was humans, who he believed also possessed immortal souls. (We can tell we have souls, he explained, in his famous dictum "*cogito, ergo sum*," because we are conscious.) His view has frequently been interpreted to mean that all other species are basically clockwork machines or robots whose cries when hurt do not imply anguish but are outputs of pure instinct. Defenders, however, say that a close reading of his own words suggests that he granted that animals had feelings, just denied that they had *thoughts*. Indeed, in letters, he spoke of animals possessing "anger, fear, hunger, and so on," "the hope of eating," (which sounds a lot like thinking to me), and even "joy." "I deny sensation to no animal, in so far as it depends on a bodily organ," he wrote. So it might be more accurate to say that Descartes thought that animals weren't really self-aware and couldn't really reason, and that they therefore didn't fall into the sphere of entities that we have moral obligations to. There are a lot of problems with this position (Is it okay to stick pins into newborn babies? They can't reason either!) but it isn't as absurd as the one often attributed to him.

For Immanuel Kant, writing in the late 1700s in what was then Prussia, all moral theory flows from what he saw as the foundational principle of ethics: you cannot treat humans as mere means to an end. You can't use humans without their consent to get something you want, even if what you want is the greater good. Kant did not, however, think this applied to non-human animals. He wrote that "animals are not self-conscious, and are there merely as a means to an end. That end is man."

But the very rational inquiry so beloved by Enlightenment thinkers began to discover the anatomical and behavioral commonalities between humans and animals, and people began to increasingly disapprove of outright torture and cruelty toward animals. In 1789, philosopher Jeremy

Bentham predicted that "the day *may* come, when the rest of the animal creation may acquire those rights which never could have been withholden from them but by the hand of tyranny." He felt that the notion that only "rational" beings should be treated ethically was absurd. He wrote, "The question is not, Can they *reason*? nor Can they *talk*? but, *Can they suffer?*"

In 1859, Charles Darwin published *On The Origin of Species*, an argument for evolution by natural selection. The scientific, cultural, and ethical ripple effects of the discovery of the common origin of humans and all other animals are arguably still playing out, with the wall imposed by Western culture between *Homo sapiens* and other species taking many generations to crumble.

In the nineteenth century, animal cruelty laws began to catch on in Britain. A major focus was on ending vivisection—surgical experiments performed on living animals. Campaigners, who were mostly Christians, saw animals as inferior to humans but argued that humans should be merciful and compassionate toward them. Victorian relationships with pets became ever closer and more like the bonds of family. Some dead cats and dogs could now expect to be mourned and buried with funeral rites, for example. ("In loving memory of our faithful little friend 'Wobbles,' August 24, 1900" reads one headstone in a pet cemetery in London's Hyde Park.)

In the West more broadly, a burgeoning interest in Asian religions, Spiritualism, Theosophy and other non-Christian ideas reopened the door to the possibility that animals might have souls in some spiritual sense—a view long rejected by Christian authorities.

But although popular attitudes toward dogs and cats and horses were becoming more overtly sentimental, many scientists disapproved. In the 20th century, an influential group of scientists known as behaviorists tried to study animals with the same rigor and rationalism that one might use to study chemical reactions. They restricted their data collection to observable and measurable data—in the case of animals, that meant recording their behavior. They might say that an experiment made a dog or monkey move away rapidly and bare its teeth, but they would never go so far as to say that the animal was afraid. They focused on instincts and on reaction to stimuli.

To speak of animals as possessing feelings or desires was considered to be anthropomorphism—to be reading animal behavior through a human lens—and thoroughly unscientific.

In the latter half of the century, this approach began to break down under the sheer obviousness of animal moods, desires, and personalities. Long-term studies, like Jane Goodall's work observing the chimpanzees at Gombe National Park in Tanzania, told the stories of individuals across their lives, and these individuals were obviously distinct from one another. Some were even jerks! The more scientists allowed themselves to see non-humans as individuals with minds, goals, and emotions, the more their studies began to show these minds busy at work attempting to achieve goals and expressing emotions. The result has been the flowering of findings in animal cognition that we toured in the last chapter.

———

In the second half of the 20th century, concern for the welfare of animals in Western countries also began to go beyond a genteel horror at wanton cruelty. Campaigners began to question deeply entrenched systems like eating meat. In 1964, Ruth Harrison popularized the term "factory farm" in her book *Animal Machines.* For many who didn't live on or near a farm, her description of modern animal husbandry was a rude awakening from the picture-book version of Bessie the cow wandering home from a field of buttercups to be milked. Readers learned about the close indoor confinement of livestock, which caused, in Harrison's view, "acute discomfort, boredom, and the actual denial of health. . . . The animal is not allowed to live before it dies," she wrote.

Peter Singer's *Animal Liberation*, published in 1975, provocatively compared human treatment of animals to white people's treatment of black slaves. His argument rests on the "principle of equality," which states that *all* sentient beings should be given equal consideration, just as all races of human beings should. That doesn't mean equal *treatment*: a chicken doesn't want or need the right to vote, for example. But it does mean that a chicken's life is just as valuable as a human's.

For Singer, acting morally means acting in a way that will result in the most good for all sentient creatures. Singer focuses on the satisfaction of individuals' *preferences* when defining the good. A dog prefers not to be kicked; an owl prefers to feed its chicks rather than see them go hungry; a whale prefers to swim to Hawai'i in the winter. Figuring out what animals *prefer* can be tricky, but we can start with the easy cases. We can assume, for example, that animals that can feel pain prefer not to feel it.

In Singer's view, every decision is a sort of math problem. Eating a roast chicken might make me feel an hour of pleasure, but the industrial chicken endured a living hell to make that pleasure possible, and since the chicken and I have equal moral worth, it is immoral to act in a way that satisfies my own preferences at the cost of a much larger and long-lasting amount of its pain. In each choice before us, the option that maximizes the preference satisfaction of all sentient creatures affected by the action is the right one.

Singer makes an interesting distinction between pain and death. Is an unexpected, instantaneous, and painless death "bad" for an animal with no specific future plans? Maybe not, from Singer's perspective. He writes, "To take the life of a being who has been hoping, planning, and working for some future goal is to deprive that being of the fulfillment of all those efforts; to take the life of a being with a mental capacity below the level needed to grasp that one is a being with a future—much less make plans for the future—cannot involve this particular kind of loss."

Singer argues for society-wide changes that were quite radical in 1975—including vegetarianism and the abolition of most animal testing—but he did so wholly using this utilitarian framework. He does not assert that animals have "rights."

Philosopher Tom Regan does. All "subjects-of-a-life" who have consciousness—or what he describes as "the mystery of a unified psychological presence"—have certain rights, including the right to "respectful treatment." They are "never to be treated as if they exist as resources for others." You might see some shades of Kant here. Again, Kant himself didn't think non-human animals were in the same category as humans, but the basic shape of Regan's argument is similar; he's just extending the circle to include sentient animals. I tend to think of animal rights advocates as "Animal Kantians." Another word for these kinds of theories that you might hear tossed around is "deontological," which more or less means "the science of duty" in Greek.

In different ways, Singer and Regan both fundamentally reject the Western idea of animals as categorically separate from and inferior to human beings—a position Singer calls "speciesist" and Regan characterizes as "human chauvinism."

Singer and Regan have both been very influential in animal ethics, but neither of them talks much about wild animals. Singer's case studies focus on domestic animals and animal testing, and he says little about wild animals except that we should leave them alone. "Once we give up our claim to 'dominion' over the other species we should stop interfering with them at all," he writes. "We should leave them alone as much as we possibly can. Having given up the role of tyrant, we should not try to play God either."

This seems at first to be inconsistent with his utilitarian framework. The world is rife with suffering and pain. Predators and parasites make many wild animals' lives miserable. They suffer accidents and disease; they starve to death. If the right action is that which maximizes the satisfaction of preferences, shouldn't we do something about all this wild pain? Singer says that while occasionally assisting wild animals in distress could be considered a good thing, any large-scale interference with the workings of wild ecosystems is likely to do "far more harm than good."

Like Singer, Regan spends much of his time focusing on domestic and captive animals. And like Singer, he believes our main ethical obligations to "wild" animals is to "let them be." "We honor the competence of animals in the wild by permitting them to use their natural abilities, even in the face of their competing needs. As a general rule, they do not need help from us in their struggle for survival, and we do not fail to discharge our duty when we choose not to lend our assistance." Indeed, Regan argues, our meddling smacks of a lack of respect. "Paternalistic intervention in animals' lives means taking measures to prevent them from pursuing what they want because, we believe, permitting them to do so will be detrimental to their interests."

The classic way to probe your own feelings about utilitarianism and deontological ethics is to picture yourself in the gut-wrenching position described in the famous "trolley problem." Imagine you are standing by some train tracks in a crowded city. A trolley has lost its brakes and is barreling down the tracks, heading for five people who are standing on the tracks, unaware of the danger they are in. They cannot hear you; you cannot warn them. But you *can* pull a switch that would divert the trolley onto another line. There's just one problem: there's one person standing on the tracks on this second line. Do you pull the switch, saving five strangers at the cost of one stranger? Peter Singer would say yes. The happiness and continued life of the five outweighs the pain and early death of the one. It's arithmetic. Kant and Regan would say no. How dare you use that one guy standing on the second set of tracks as a tool to save the lives of the others?

Now here's a twist. What if instead of pulling a switch to save the five, you are standing on a footbridge above the tracks and would have to physically shove one person off the bridge and onto the tracks to stop the trolley? Does your answer change? Singer's and Regan's don't.

―――――――

In response to both Singer and Regan, a number of feminist philosophers reject both of their approaches as too rule-based and insufficiently attuned to the specific relationships between thinking entities. When I read Singer's and Regan's books, I, too, had a sense that there was something very stereotypically "masculine" about their approaches. Their principles—to maximize satisfaction of preferences and to never treat an individual as a means to an end—seem very simple, fair, and almost math-like. But they explode into practically impossible duties very quickly. Singer's rules seem to require that we dedicate all of our resources, apart from the bare minimum required for living, to reducing global suffering. Regan's moral principles would seem to rule out *ever* harming or killing any sentient animal in any context other than true self-defense. And they both claim that moral judgements should be made with our rational, logical minds, not our emotions.

For example, Regan feels that our intuitions about what is right and what is wrong are valuable, but only our "reflective intuitions," which he defines as "those moral beliefs that we hold *after* we have made a conscientious effort . . . to think about our beliefs coolly, rationally, impartially, with conceptual clarity, and with as much relevant information as we can reasonably acquire." In other words, these intuitions must be emotion-free.

In contrast, philosopher Lori Gruen sees emotions like compassion and empathy as central to ethical decision-making. Gruen says many feel "alienated" by theories that ask people to ignore the particular complexities of any given situation and to rely on sweeping rules and abstract rationality. Instead, Gruen argues for approaches that more closely resemble how humans actually do make ethical decisions—using both our rational and emotional faculties, taking into account the relationship between our own multifaceted selves and the animal or animals we are considering.

Gruen's approach can be categorized as "moral particularism"—a school of thought that holds that the most moral people aren't "principled." Rather than living by maxims or rules or commandments, the most moral way to act is to pay careful attention to all the features of a particular case, develop defensible reasons for acting in a certain way, and then act. Your reasons, though, do not have to be applied consistently from case to case, nor do you have to involve appeals to overarching moral principles. Each case is different, after all. This approach leaves lots of room for context, relationships, and feelings to matter.

Gruen suggests that our moral decision-making should emerge from our "entangled empathy" with animals. She defines "entangled empathy" as "a type of caring perception focused on attending to another's experience of well-being. An experiential process involving a blend of emotion and cognition in which we recognize we are in relationships with others and are called upon to be responsive and responsible in these relationships by attending to another's needs, interests, desires, vulnerabilities, hopes, and sensitivities."

The idea that this kind of approach is "feminine" as opposed to the "masculine" principle-based approaches of Singer and Regan strikes Gruen as "unfortunate" because "it further entrenche[s] stereotypical gender roles and seem[s] to preclude the idea that men are caring." Obviously, anyone can find empathy for non-humans.

Indeed, whether you believe it to be better or worse than the utilitarian or "animal Kantian/deontological" approaches, the vast majority of people actually make their daily ethical decisions in a way that could be described as particularist. Cognitive scientists who study decision-making have identified at least two kinds of processes happening in our brains when we make any kind of decision at all, moral or not: a quicker, automatic emotional process and a slower, conscious, more rational process. Because the emotional response is faster, we often end up using our conscious thinking process to rationalize the decisions we've already made.

This might explain why people confronted with the trolley problem often give a split decision. Most people will pull the switch, suggesting that they think a bit more like Peter Singer. But those same people are often unwilling to push the person to his or her death, now seeming to side with Regan and Kant. It may be that it is just too emotionally horrible to imagine creeping up behind a person (I always imagine him as a man in an orange Hawaiian shirt and a straw hat, licking a vanilla ice-cream cone), and physically shoving him—your muscles working against him and your hands feeling the heat of his body for a moment before he tips irretrievably into oblivion, screaming in confusion and terror as he falls. Indeed, when you pop people in a brain scanner and pose these dilemmas, you can see the emotional charge of this "shoving a person off the footbridge" case light up areas associated with emotion whereas the thinking in the switch-pulling case seems not to involve these areas.

I see experiments like these as both a reality check and a warning. They are a reality check in the sense that they describe what's actually going on: human animals trying to solve ethical problems using the squishy biological brains bequeathed to us by the amoral process of evolution and honed by survival, not by rationality or morality. We are not working with tools designed for ethical decision-making.

These experiments are also a warning, because they remind us that emotions do play a major role in our ethical decision-making and that insofar as we feel emotions like fear or disgust about people or animals that our culture has trained us to see as scary or gross, we may not treat them with the same respect or afford them the same rights as those people or animals we've been trained to admire. In other words, biased thinking is inevitable. We must learn how to look for it and fight it.

When a journalist at the *Atlantic* asked one of the scientists who study moral cognition, Joshua Greene at Harvard, how to know "when we're engaged in genuine moral reasoning and not mere rationalization of our emotions," he said we have to feel a sense of effort. "[D]o you find yourself taking seriously conclusions that on a gut level you don't like? Are you putting up any kind of fight with your gut reactions? I think that's the clearest indication that you are actually thinking it through as opposed to just justifying your gut reactions," he said.

In some ways, choosing a moral principle and then applying it consistently in every case, no matter how you *feel* about it ensures that you aren't letting your emotional mind call the shots and can protect you from your biases. But philosophers like Gruen say that these principles simply won't always give you the right answers. They are blunt instruments, too divorced from context.

Gruen and others see relationships of care as central to ethical decision-making, not as irrelevant externalities. Under these theories, we absolutely have different and much more elaborate obligations to a pet than we do to a "wild" animal because of the different nature of our relationship. We might not even have any obligations to a "wild" animal. Or, we may not feel obligations of care for "wild" animals in general, but if we stumble across an injured animal while on a hike, we might feel bad for it and feel a sudden duty of care that springs out of the situation.

Operating on the basis of empathy could also be tricky in some contexts because predators make their living by visiting pain and death upon their prey. With which side of this interaction should we empathize?

One final ethical approach to consider is called "virtue ethics." This framework does not offer rules or principles for deciding what the ethically correct thing to do is in a given situation, and in that way, it is a particularist theory, like Gruen's entangled empathy. But rather than focus on emotions and relationships and guides, virtue ethics focuses on specific human virtues.

As applied to non-human animals, a key virtue is compassion or "respectful love," according to philosopher Rosalind Hursthouse. Respectful loves combines caring about the welfare of others with remaining "mindful of others' rights to make their own choices." In environmental virtue ethics, though, the key virtue is humility. Through this lens, it is arrogant to treat the non-human world as a pile of resources there for the taking or as a system we can manipulate at will; humility means seeing oneself as just a single individual among many and humans as just one species among many. It means acting with restraint and thoughtfulness for the needs of others. Often, the virtues of compassion and humility will suggest the same course of action. But in some cases, compassion may suggest action while humility may suggest inaction—or even vice versa.

———

Today, a few lawyers are bringing back the idea of some animals as "persons," seeking to legally erase the line between humans and other animals. The Non human Rights Project's goal is to "change the common law status of great apes, elephants, dolphins, and whales from mere 'things,' which lack the capacity to possess any legal right, to 'legal persons,' who possess such fundamental rights as bodily liberty and bodily integrity." If the idea of animals as legal persons sounds far-fetched, just remember that in the United States, corporations are legal persons. So membership in the species *Homo sapiens* is clearly not part of the legal definition.

To try to achieve their goal, the Non human Rights Project represents apes and elephants that are in captivity, and sues their "captors" under habeas corpus, the basic right not to be detained or held without being charged with a crime. The judges who are assigned their cases have to make up their own minds about whether the clients before them qualify as persons. By continually filing new cases in new jurisdictions, the lawyers at the Non human Rights Project hope to eventually get a sympathetic judge who will rule that Hercules and Leo (former research chimps) or Happy (an elephant kept all by herself at the Bronx Zoo) are people. That would set a powerful legal precedent.

In 2018, a team of lawyers led by attorney Steven Wise came close in the case of Tommy, a chimp kept all alone in a bare cage in a shed by his "owner." After a lower court threw the case out, the team went to the New York Court of Appeals—twice. The second time, the court denied the motion for permission to appeal on the grounds that it had already denied permission once. But at least one judge, Eugene Fahey, was beginning to have second thoughts: "To treat a chimpanzee as if he or she had no right to liberty protected by habeas corpus is to regard the chimpanzee as entirely lacking independent worth, as a mere resource for human use, a thing the value of which consists exclusively in its usefulness to others. Instead, we should consider whether a chimpanzee is an individual with inherent value who has the right to be treated with respect. . . . In the interval since we first denied leave to the Non-human Rights Project, I have struggled with whether this was the right decision. Although I concur in the Court's decision to deny leave to appeal now, I continue to question whether the Court was right to deny leave in the first instance."

Even as I find myself increasingly willing to think about animals as persons, I admit that there does seem to be something different about humans. Unlike a deer or a bat, we worry about the ethics of our actions. We can think abstractly about what makes an action right or wrong. Now, it isn't fair to say that animals have no concept of care, kindness, or empathy. There are many examples of animals helping and caring for one another. And some non-human animals clearly understand fairness. One experiment looking at this question ended with a capuchin monkey throwing a cucumber in the face of a researcher in righteous indignation when the monkey next door was rewarded with a much more desirable grape for the same task. But despite these glimmerings of animal morality, it would seem cruel to punish animals for doing "bad" things, as medieval Europeans occasionally did. We don't expect the lion to refrain from eating the lamb because the lamb objects. In ethical parlance, humans are "moral agents," who have obligations to do the right thing, while animals are simply "moral subjects"—entities toward which we may have ethical obligations.

Singer, Regan, and Gruen all seem to believe we have limited obligations to animals that are neither our livestock, nor our pets, nor our captives. We should simply leave them be. But "wild" animals in the twenty-first century are not truly independent of humanity. In fact, today, I am not sure there are any "wild animals" left.

4

Between Dog and Wolf

In July 2013 a young black wolf was walking near Ione, Washington, in the state's rural northeast corner. She stepped in a leghold trap—the kind without pointy teeth—and was caught. Some curious calves found her and came close to investigate—two free-roaming domestic animals checking out a captive wild one. The calves' mothers arrived and were concerned about the wolf, so they began bluff-charging her, running toward her aggressively then veering away at the last moment.

Humans then approached, chased away the cows, and jabbed the wolf in the flank with a tranquilizer-filled syringe at the end of a long pole. After a furious, snarling initial response, the wolf began to drift away. When she was fully unconscious, the state biologists moved in to get to work.

The team estimated her weight as 68 pounds and her age as 15 months. They fit her with a GPS collar, placed an electronic ID chip under her skin and clipped a bright tag onto her ear bearing her ID number: 47.

The biologists backed off while 47 slept off the tranquilizer. After she woke up, she and her companion—another female—continued along their way. At sexual maturity, wolves tend to leave home and hit the road, looking for a mate and territory of their own. Often, siblings will travel together for a time, so 47's companion might well have been her sister. One day, they encountered a giant Akbash livestock guard dog. This fluffy white breed is supposed to run off wolves; that's its job. But for whatever reason, these wolves and this dog did not see each other as enemies. They began palling around with one another.

Wolves are, for many people, the living embodiment of wildness. Dogs, on the other hand, are the non-human animal we most frequently consider a part of the family—"man's best friend," the apogee of domestication. And yet, the two creatures can and do mate and have fertile offspring. The gray wolf's scientific name is *Canis lupus*. The dog's scientific name is *Canis lupus familiaris*. That is to say, they are the same species.

Washington State officials worked with the family who owned the dog to install a nine-foot fence around its yard. Still, the female wolves hung around and were felt to be coaxing the dog out. Finally, the 90-pound Akbash managed to scale the fence and, in the words of Donny Martorello, the carnivore section manager for the Washington Department of Fish and Wildlife, "decided to be a wolf for a month."

State officials did not want the dog to impregnate either of the female wolves. For one thing, the genetic integrity of the wolves was at stake. "Genetic integrity" is a phrase commonly used in the conservation biology literature to evoke the value of a genome that is not tainted with the genes of related species or subspecies. (There already are some dog genes in the wild wolf population. The coat color of North American black wolves derives from a trait that first evolved in the domestic dog. But the trait passed to wolves so many millennia ago that those genes have been given honorary wolf status, so to speak.)

In addition, dogs are considered property, whereas wolves were an endangered species statewide in 2013. The rules and regulations governing them were wildly different, and neither set of rules is clear about how to handle hybrids.

In essence, the three animals were outlaws. Local ranchers named the two females Thelma and Louise. The GPS collar on Louise (also known as 47) made it a relatively simple matter to find the two wolves and the dog and, using helicopters, dart the wolves with tranquilizers. The wolves were examined while the dog stood by, furiously barking and guarding the females. Thelma was found to be pregnant.

After consultation with the state wolf advisory group, officials recaptured the pregnant wolf, again by helicopter, and spayed her—that is, removed not only her embryos but also all her reproductive parts. Martorello explained that full spaying was deemed to be safer and easier than giving her an abortion. The dog was returned to its owners. The two wolves continued to frequent the area. A few months later, Thelma was struck by a car and killed.

Louise returned alone to the house where the sheepdog lived. Efforts to drive her away from the dog's enclosure proved unsuccessful. "She thinks that male dog is a pack member," Martorello told me at the time. "Over the summer and early fall we did everything to haze the animal: rubber bullets, chasing it, trying to catch it up. We couldn't do it. She is still a wolf and she is wily."

In February 2015, Louise was finally cornered. She was knocked out by a tranquilizer dart shot from a helicopter and moved to Wolf Haven International, a sanctuary for pet wolves and wolf dogs near Olympia, Washington. Here she will spend the rest of her life in captivity. She has been "partially altered" so she can never breed; she is placed with a similarly altered male wolf dog for company. Her name was changed yet again, this time to Ione.

Wendy Spencer, the director of operations at Wolf Haven, told me that the facility usually does not take wolves from the wild, believing that even the generously sized enclosures and limited human contact they provide is not an acceptable quality of life for a really wild wolf. But 47/Louise/Ione had already become used to humans. "Based on all the reports we have from the field about how habituated she was, she is probably a good candidate for life in captivity," she says. "It was that or be shot."

Today, Ione seems relatively content. Pamela Maciel Cabañas, sanctuary co-manager, gave me an update in 2020, saying Ione "continues to live a peaceful life in the off-public area of our sanctuary. She shares a heavily vegetated enclosure with her male companion, wolf dog Luca, with whom she seems very bonded. Animal care staff often observe them being playful with each other or sharing a large raw-meat meal. Ione has the wary temperament typical of wolves, and usually when there is human activity in the area she chooses to keep out of sight by retreating into the thick bushes using a complex system of tunnels that she and Luca have created over the years. In contrast, when the staff truck loaded with food approaches their enclosure, it's not uncommon to see Ione join Luca in a series of displays of excitement, which can include jumps high in the air."

Washington State had a mandate to protect the gray wolf as an endangered species; hybridization would have created animals of uncertain status that would entail uncertain obligations, possibly including the obligation to kill the pups as a threat to the genetic integrity of nearby wolf packs or as vermin. A wolf dog in the wild is worse than useless; it is the seed of chaos, the harbinger of the "hybrid swarm," in which all order breaks down and wolves lose their value.

Italian wildlife managers have had to cope with their own hybrid swarm—a pattern of matings between dogs and wolves so pervasive and long-standing that most Italian wolves are wolf dogs of varying percentage. Genetic studies suggest many of the hybridization events took place generations back as the Italian wolf population expanded into new areas, which suggests that such events may become increasingly common in Northern Europe and North America, where wolf populations continue to expand into new territories filled with humans and dogs.

Luigi Boitani has studied Italian wolves for decades, and says that, as you might expect, there is no black-and-white prescription for dealing with hybrids. "Of course, you do not want to kill all the black wolves of North America, that would be ridiculous," he says. "But you would probably want to intervene if you saw a wolf mating with a dog in your garden. Between these two extremes, there is a continuum. And it is very hard to draw a line."

Boitani cheerfully admits that wolves and dogs are the same species, that they are very difficult to tell apart, even with genetic tools, and that hybrids do as well or better than "pure" wolves on the human fringes where they live in Italy. In 1975, he radio-collared three free-ranging hybrids and followed them for a year. "They behaved 100 percent like wolves," he says. They defended home ranges; they killed livestock. The only difference was that the male dogs made lousy fathers, failing to bring bellies full of meat back to the den to regurgitate for the pups like wolf dads do. Nevertheless, "a she-wolf that mated with a big sheep-guarding dog" that Boitani observed raised six pups all alone. All dispersed and went on to lead wolfy lives.

Despite the fact that these hybrids more or less act like wolves, he still thinks the population should be managed to make it "wolfier." How? You just kill the ones that look like dogs, Boitani says. "If they look like wolves, if they behave like wolves, if they have an ecological function like wolves, then leave them. On the other hand, if you have a wolf that looks like a dog, then I would have a problem. I would like to have around a wolf that looks like the wolf and is not polluted with our dogs."

Here, the recommended management strategy is to keep the wolves looking like wolves. To Boitani, the wolves' aesthetics and the preservation of their ecological niche are the object; the preservation of purely "wild" DNA and the freedom of individual wolves, less so.

Western culture famously loves categories and especially dualities. Man and nature. East and West. Wild and tame. The way we deal with the prospect of a wolf dog in the wild shows our deep displeasure at any blurring or breaking of the binary: As the symbol of wildness, wolves must be absolutely free of human influence—pure. By allowing them to mate with dogs, we sully this wild purity with grubby DNA from the dog, the icon of domestication.

Wolves as symbols of wildness are so culturally important that we humans will go to great lengths to protect the species' purity, even if doing so involves restricting the freedom of actual animals. Wildness is often defined as that which is not controlled, but paradoxically, in order to protect the "wildness" of the wolf gene pool, individual wolves must be controlled.

Conservationists, in particular, seem to value what we might call "wild DNA"—a genome untainted with domesticated genes—above the behavioral or functional autonomy of actual animals. The very existence of the fields of "wildlife management" and "wilderness management" suggests that we are surprisingly comfortable exerting some kinds of control over organisms and places we designate as "wild." But then, human attitudes toward the wild have never been simple.

————

As far back as written stories go, we see both awe and fear, delight and disgust in the wild world. We often see the urge to conquer and dominate, but we also see, like a transcendent counterpoint, an urge to be swallowed up by the wild, to merge with it.

The Epic of Gilgamesh, one of our oldest written narratives, chronicles the great city builder Gilgamesh's friendship with a "wild man," Enkidu, who, when we meet him, is living contentedly without human contact, "[c]oated in hair like the god of the animals," grazing at the water hole with a herd of gazelles, "his heart delighting with the beasts in the water." Gilgamesh forcibly humanizes Enkidu by sending Shamhat, a priestess, to seduce him. After many days of lovemaking, Enkidu is spurned by the wild animals. He gets a haircut, eats bread and ale, and returns with Shamhat to the city. After an epic street fight, he becomes Gilgamesh's best friend and together they hit the road, slay the forest god Humbaba, and take his tusks as a prize. Afterward, they fell the biggest and best tree to take back to Uruk to make a fancy door. It is hard not to read this deforestation parable as a metaphor for the rise of humans as an ecological force—the triumphant expansion of urban and agricultural lands into formerly wild places. But even as Gilgamesh and Enkidu enter the forest, kitted out with axes and daggers, they pause for a moment to admire it, "gazing at the lofty cedars" and noting that "its shade was sweet and full of delight."

Something about this moment reminds me of historical photos of nineteenth-century lumberjacks standing next to enormous trees, many halfway sawn-through. These images, which I saw often growing up in the Pacific Northwest, speak to the loggers' pride in being able to achieve such huge tasks as well as a real awe at the sheer scale of these trees. Loggers tend to have a real love for trees in the way that many hunters truly respect their quarry.

Appreciation and awe are woven into humans' most ancient feelings about wilderness. In the Bible, wilderness is a barren, life-threatening place, but it is also a place of revelation. In Genesis, Eden is a garden, but one that needs no tending, a kind of primeval wilderness with the sting of death removed. And in Christian accounts, the expulsion from Eden and the adoption of an agricultural way of life was a tragedy.

Consistent throughout is the same duality, the belief that humans—and the agricultural landscapes and cities they create—exist in a separate category from all other species and the landscapes they create.

In contemporary Western thought, this dualism often takes on a charged moral color. More than just different from the wild, humans are seen as capable of destroying the wild with their touch. In this framing, any human influence on nature is bad. All changes that flow from human action are degradation no matter what they are.

This generalization is understandable. Humans have made so many changes that we now regret. We look at our oceans swirling with plastic, our shrinking old forests, the dying eyes of one of the last 10 vaquita porpoises caught in a net, and we understandably feel strongly that we ruin everything. This view was memorably captured in March 2020 by a viral tweet that quickly became meme shorthand for taking a grim, misanthropic pleasure in the Covid-19 pandemic.

Wow . . . Earth is recovering
—Air pollution is slowing down
—Water pollution is clearing up
—Natural wildlife returning home
Coronavirus is Earth's vaccine
We're the virus.

This idea that humans are like a disease is not new. I remember attending a scholarly talk at a 2009 meeting of the Society for Conservation Biology in Flagstaff, Arizona, in which a professor compared cities to cancers, in both their physical growth patterns and in their moral value.

The flip side of the idea that humans are bad by definition is that nature is good by definition. Thus we see the word "natural" splashed across our breakfast cereal, our shampoo, our dish soap. Expectant mothers opt for "natural" births, eschewing hospitals and anesthetics; shoppers flock to "natural" food stores, as if mainstream grocery stores sell toy broccoli and plastic steaks.

"Wild" and "natural," "wilderness" and "nature"—these terms are often used interchangeably when talking about things, animals, and places. There have been whole books written about the meaning of these words. I wrote one of them myself.

In United Sates law, federally designated wilderness is famously defined as "an area where the earth and its community of life are untrammeled by man, where man himself is a visitor who does not remain." One environmental ethics text defines natural like this: "Something is natural to the extent that it is independent of human design, control, and impacts." Definitions like this start with a basic assumption that human beings are not part of nature. They assume, in fact, that humans are the opposite of nature, that our influence makes a thing less wild or natural. And I simply reject this premise.

After many years, I have come to see the concepts of wilderness and nature as not just unscientific but damaging. Firstly, all organisms alive today are influenced by humans. Secondly, we humans are deeply influenced by the plants and animals we evolved with; we are part of "nature," too. Thirdly, "wilderness" rhetoric has long been used to justify denying land rights to Indigenous people and to erase their long histories. And finally, thinking of nature and humans as incompatible makes it impossible to revive or discover ways of working with and within nature for the common good. I still have hope for the word "wild," but as a term for *autonomous*, not un-human.

———

Let's start with that first point. All Earth's species have been influenced by humanity's changes to the environment. Even animals that have never seen a human are changed by our species, genome deep. In a sense we are only beginning to understand, "wild" animals today, and the ecosystems they live in, are all party human creations.

All species that regularly interact act as selection pressures for one another, shaping each other's evolution. Natural selection favors organisms that thrive in their environment, and the environment is as much the living species in a place as it is nonliving factors like climate. So like all animals on Earth, our species has been affecting other species for its entire run. For much of our history, our line hasn't had a particularly outsized influence. We may have been a minor factor in the evolution of some predators by being tricky prey and we certainly molded the evolution of our direct parasites, like lice. *Pediculus humanus* can thrive on no other animal. But in general, we were just one animal among many, all creating evolutionary pressures for each other, in a complex web of pushes and pulls.

Our career as "super influencers" began in earnest relatively recently, evolutionarily speaking. There are hints that our ancestors in the Pliocene were so good at stealing food from large carnivores that we helped drive these great beasts extinct. One study says that the timing of extinctions of "species of bears, saber-toothed cats, and giant species of martens, otters and civets" in East Africa correlates better with increases in our ancestors' brainpower than with climate changes. (Though their model also suggested that a decline in forest cover could also be an important causal factor.) In a time when competition for prey was fierce, being driven away from a kill by stick-wielding hominins might have been the difference between life and death.

By the late Pleistocene, though, *Homo sapiens* had arrived, and we weren't just stealing food from better hunters; we had become the best hunters on Earth. A wave of extinctions of animals over about 100 pounds (known to researchers as "megafauna") followed humans as they migrated around the globe. There are a huge number of these extinctions, and ultimately, each one has a slightly different causal story. Climate changes are likely to have contributed to many of them. But there are few scientists now who argue that human hunting—both of the extinct animals themselves and the prey they depended on—wasn't a factor in at least some of these extinctions.

North America lost more than 70 percent of its megafauna in the Pleistocene, including the massive dire wolf, the American lion, giant ground sloths, a species of mountain goat, saber-toothed cats, the giant short-faced bear, mammoths, and mastodons. It is impossible to fully imagine the continent as it was when inhabited by these giants. Over my fireplace, I have a painting by my brother Alex of the skeleton of a short-faced bear, with the skeleton of a human being beside it, for scale. When standing on its hind legs, this bear was 12 feet tall. It could run up to 40 miles per hour. It ate meat. Probably, it ate us. When I gaze up at it, I try to picture what kind of culture, what kind of psychology, what kind of person could cope with living in a world where such raw fierceness was a real existential threat. I simply cannot.

Between 15,000 and 10,000 years ago, all these animals disappeared. Since they, like all creatures, influenced the species in their ecosystems, their landscapes transformed in their absence. Paleoecologists can look back in time by extracting long muddy cores from lake beds. The cores are like thousands of layered time capsules, each capturing one period of time through the sediment that gradually settled on the bottom of the lakes. Scientists look in each layer for charcoal from wildfires, pollen grains, and spores of the fungus *Sporormiella*, which live in herbivore dung. These studies have shown, for example, that when mastodons, mammoths, giant beavers, giant sloths, and a giant moose relative (*Cervalces*) declined in numbers across what is now the area from Indiana to New York—as evidenced by a crash in *Sporormiella* spores— broadleaf trees exploded in numbers, overtaking spruce as the most frequent source of pollen. As anyone who has tried to garden in a place with deer knows, herbivory can be a powerful force, keeping palatable species down while spiky or toxic plants flourish. The decline of the megafauna had an effect not unlike erecting a deer fence around the entire continent. Less grazing meant more plants, which also meant more fuel for fire, and researchers saw more charcoal in the sediment layers laid down after the extinctions as well.

In Australia, humans arrived much earlier, some 65,000 years ago. There was a pulse of extinction there following the arrival of humans, and, as in North America, a transformation of many landscapes. Areas of the Outback that are now nearly treeless were once "a mosaic of woodland, shrubland, and grassland, with a high proportion of plants with palatable leaves and fleshy fruits," according to one analysis. As giant relatives of kangaroos and wombats and huge flightless birds disappeared, plant matter built up and then burned, starting a cycle of wildfires in some places that favored the tough, fire-adapted species now common across the arid parts of Australia.

The first people in New Zealand arrived less than 1,000 years ago—so recently that their lost megafauna is just barely gone. At Yale's Peabody Museum of Natural History, I saw a full skeleton of a Moa—a family of gargantuan birds up to 10 feet tall that, like so many others, was likely ushered into extinction by human hunting. I looked down at the card and the hairs on the back of my neck stood up as I read the word "subfossil." I realized I wasn't looking at hard minerals that had taken the place of bones, as in your typical dinosaur fossil. I was looking at this bird's *actual bones*—as greasy and real as the bones left over after turkey dinner. That's how recently we lost the nine species of Moa.

But even though the bodies are barely cold, the land has already responded to the change. Without the big herbivorous Moa keeping forest open enough for sunlight to reach the ground, at least in some places, the forest closed in and shade-adapted species took over.

All across the world, human hunting likely contributed to megafaunal extinctions, which in turn led to significant changes in ecosystems. These changes presented new challenges and opportunities for the surviving animals, influencing their behavior and, over the generations, shaping their evolution.

There were likely some secondary extinctions—certainly of parasite species, perhaps also species of dung beetle that relied on the not insignificant dung heaps left by some of these massive creatures. Similarly, scavengers were in trouble, with fewer giant carcasses to feast on. In North America, we lost seven entire genera of vultures, and the California condor only survived by taking advantage of beached whales and other marine mammals.

Other species had to change their behavior to survive. The common vampire bat was likely hit hard by megafauna extinctions in its home range in the Americas. This bat lands on sleeping mammals, neatly slices open their skin, and laps up their blood. With fewer large animals to sip from, life was tough, but the bat adapted by switching over to humans and their live-stock. Today, pigs, horses, and cattle are their preferred prey.

Many large predators, such as dire wolves and saber-toothed cats, went extinct during this time, but others evolved to thrive in the new normal. In South America, jaguars shrank to scale with the average size of their new prey. Instead of gorging on wild horses and camels and ground sloths, they learned to make do with capybaras and giant anteaters. Coyotes shrank too, and used their legendary cleverness to survive in the new, smaller world they found themselves living in.

There's not a lot of research on specific animal adaptations to the extinc-tion wave. As one scientific paper put it, "To date, the influence of the terminal Pleistocene extinction on the surviving small and medium-sized mammals has been largely ignored." That paper looked at a serious grave site: Hall's Cave in the Texas Hill Country, where bones have been building up for 22,000 years. Their analysis found that the extinction pulse knocked out 80 percent of the large-bodied herbivores and 20 percent of the apex predators in the area. Afterward, animals were, on average, smaller and there were fewer grazers.

The paper's authors noted another interesting finding: Today's apex predators—jaguars, mountain lions, wolves, grizzly bears—used to be the little guys making a living in the shadows of even more imposing specimens: the saber-tooth and scimitar-toothed cats, dire wolves, and the short-faced bear. These extinct "hyper carnivores" usually specialized in just one or a few prey species. In their wake, we are left with "apex" carnivores that grew up being scrappy opportunists. They can scavenge, eat plants, switch prey, and make other behavioral choices. That flexibility is no doubt serving them well now as they struggle to hang on in the twenty-first century.

In Africa and Eurasia, the effects of human hunting are more conten-tious, and some megafauna likely had time to evolve responses to the increasingly clever apes to avoid extinction—but that doesn't mean our influence was not as profound. Indeed, if such animals as African elephants, rhinos, hippos, bison, elk, and water buffalo survived our rise as efficient hunters by evolving new defenses—such as better hearing, sharper horns, or defensive behaviors—that fact only supports the argument that human influence is already woven into the genomes (and lives) of these species.

Humans didn't only shape other species by contributing to extinctions. In other cases, we evolved mutually beneficial relationships with non-humans that changed both parties.

Robin Wall Kimmerer is an ecologist and enrolled member of the Citizen Potawatomi Nation. She's also the director of the Center for Native Peoples and the Environment at the State University of New York College of Environmental Science and Forestry in Syracuse. In her book *Braiding Sweetgrass*, she writes about how harvesting fragrant sweetgrass to make baskets encourages its growth—and how unharvested patches of the plant wither away and die. "With a long, long history of cultural use, sweetgrass has apparently become dependent on humans to create the 'disturbance' that stimulates its compensatory growth," she writes. M. Kat Anderson, an ethnobotanist at the University of California, Davis, writes that "It is highly likely that over centuries or perhaps millennia of indigenous management, certain plant communities came to *require* human tending and use for their continued fertility and renewal." In other cases, humans directed the evolution of food species by selective harvesting and by replanting seeds of individuals with desired characteristics.

Indigenous North Americans moved plants around intentionally, bringing fan palms to the Sonoran Desert for shade and fruit, for example. The Kumeyaay, who live in and around San Diego, cultivated manzanita, ceanothus, and wild roses and actively extended the range of plums, agave, yucca, sage, and mesquite. They brought prickly pear cactus from the desert to the seashore and planted orchards of oaks and fields of grain. "Careful study of Spanish accounts indicates that little or no natural landscape existed," anthropologist Florence Shipek writes. "Instead they describe grass, oak-park grasslands, limited chaparral, and areas with plants so even and regular they looked planted, all the product of human management to provide food." The Kumeyaay's neighbors and trading partners, the Chumash, may have brought gray foxes to the Channel Islands off the coast of California—which then evolved into an entirely new species, the miniature island fox.

Many Indigenous peoples use fire to manage both the land and its animals. Fire can keep down dry fuels, reducing the chance of a destructive, out-of-control fire later in the season. Fires also stimulate new plant growth, which is the most nourishing to many herbivores. So the practice both feeds wild animals and attracts them to be hunted. Many American prairies, meadows, and grasslands that colonizers assumed were "natural" were in fact intentionally maintained with fire.

Indigenous land management would have influenced the animals that lived in and around these ecosystems. More directly, humans altered the evolutionary trajectory of some wild animals so thoroughly that they ceased to be wild. One theory about how dogs came to be domesticated focuses on wolves and humans as ancient hunting partners. Indigenous peoples from North America, Siberia, and Eastern Asia, all tell stories of cooperative, friendly relations with wolves.

Indeed, it is possible that the reason there aren't more examples of mutualisms between humans and wild animals is that, in such cases, we have become so thoroughly intertwined that we call our "partners" something else: domesticates. Dogs are, in this obverse framing, just the subset of wolves who have mutualistic relations with humans.

Today, even the wildest of wild animals are not only influenced by all those millennia of human-caused changes, they are continuing to adapt to our ever-changing ways. Wild animals make their own choices about what to do every day—in that sense they are free. But their bodies and minds evolved in a world profoundly influenced by humans, and all too often, their daily choices involve navigating a world that has been rearranged for human needs and desires.

Elephants are losing their tusks as poachers kill the big tuskers before they can reproduce. Animals we hunt and fish for intensively are becoming smaller-bodied as they evolve to reproduce faster. In Alberta, conservationists tried to train grizzlies to keep away from roads so they wouldn't get hit by cars—but the bears were too smart for them and only ran away from white trucks, like those driven by the wildlife managers.

As for the "wild" animals in our cities and suburbs, they have thoroughly adapted to our world. Animals that communicate by sound, including birds, frogs, and toads, have shifted the pitch of their calls and songs to be heard above the noises of cities and traffic. White storks in Spain stopped migrating when they realized they could just eat trash at the dump and introduced crayfish. Crows in Sendai, Japan, wait for traffic lights to turn red, put walnuts in front of the tires of idling cars, then pick the meat out of the nuts once the cars have run them over and cracked their shells. Rats, pigeons, house sparrows, and other commensal organisms have so fully adapted to human beings that they now depend on humanity.

House sparrows, for example, are completely at home in the human world, nesting in parking lots and outdoor light fixtures. They have followed humans across the globe, eating our spilled grain, insects stuck in the grilles of our cars, and our garden plants. There are upward of 540 million of them—all living right under our noses. House sparrows, like other commensals, are clever and adaptable. In New Zealand, they have mastered automatic sliding doors, triggering the mechanisms to get into lunchrooms and cafés in buildings. They have even learned to incorporate cigarette butts into their nests as a form of pest control, since the nicotine repels parasitic mites.

Climate change is altering the ranges, annual cycles, and behavior of untold numbers of species. Great tits in the United Kingdom are laying eggs two weeks earlier, tracking the changing schedules of the caterpillars their nestlings eat. The caterpillars are, in turn, tracking the bloom time of trees. Some caterpillars in the United States are even evolving to lay their eggs on new plants—"weeds" from across the sea. These new relationships between new and native species are knitting together novel ecosystems around the world.

Animals are moving toward the poles and upward in elevation as the climate warms. In mainland Britain, a study of invertebrates, including bees, butterflies, grasshoppers, and spiders, showed them moving north at an average rate of 1.2 miles a year. Over half of plant and animal species in temperate North America have seen their ranges contract at the hotter edge of the range, expand at the cooler edge, or both. Not all animals move, though, or move the way you might predict. Researchers seem to be uncovering more ability to adapt in place than they expected, which is encouraging.

When animals move uphill, the cone shape of mountains means they get squeezed into smaller and smaller areas. Studies are showing this "escalator effect" in real time. On one ridge that rises almost 1,000 feet out of the Amazon rainforest in Peru, for example, scientists surveyed for birds in 1985 and in 2017. In the 32 years they were away, the local average temperature had increased by about 0.42°C and birds had moved upward an average of 40 meters. Six species formerly only found at the ridge top—including the crested quetzal, the Andean motmot, and the hazel-fronted pygmy tyrant— seemed to have run out of options. They could not be found at all in 2017.

Animals tend to evolve smaller bodies in hotter temperatures, perhaps because they are less prone to overheating with a larger surface-area-to-volume ratio. Scientists have caught a population of a South African bird, the mountain wagtail, shrinking by 0.035g per year as its habitat warms. The American lobster and Atlantic cod have also shrunk as the northwest Atlantic Ocean has warmed.

Climate change is increasing plant growth in some ecosystems, creating more food for animals, but drought is setting up starvation scenarios and triggering catastrophic wildfires in others. More than any other human influence on Earth, climate change is global in scope, affecting all life, all ecosystems.

Few argue that *any* level of human influence disqualifies a place from being "nature." Such a purist stance would imply that "nature" ended when anthropogenic (human-caused) climate change began. For many, "nature" is a property that comes in degrees, and one can speak of places as being more or less natural. But even in this sense, there is no real room in the concept for us to imagine human-non-human relationships that are not destructive, since human influence by definition decreases naturalness.

More troublingly, not all humans are seen as separate from "nature." The way people use "wilderness" in particular constantly perpetuates the colonialist myth that Indigenous people had no agency and could not modify, manage, or influence the landscapes around them. To this day, professional ecologists measure the "ecological integrity" of North American landscapes by comparing their current state to the "natural range of variability," which is typically defined as whatever it looked like in the three or four hundred years just before Europeans showed up.

And yet, Indigenous land management was in many places creating those "natural" states: prescribed burns maintained prairies and grasslands; hunting determined populations of prey species; harvest and replanting shifted the ranges and abundances of some plant species; agriculture domesticated others. When Christopher Columbus arrived, a good 10 percent of the land area of the Americas was being intensively farmed or used for settlements. Where colonizers saw this management in action, they often failed to recognize it, because of its different forms and because of their preconceived notions about the Native people.

But in many places, colonizers never even saw the management when it was up and running. Because of warfare, enslavement, hunger and especially the swift and massively lethal spread of measles, smallpox, influenza, and the bubonic plague from European colonizers to Indigenous people, 90 percent of the pre-Columbian population died by the beginning of the 1600s—56 million people—an absolute cataclysm. And these deaths occurred before white people even saw most of the Americas. The changes to the land were massive. As the forest reclaimed fields and villages, new trees sucked up enough carbon dioxide to reduce the greenhouse effect and lower the Earth's temperature by 0.15°C. In many places, it was this apocalyptic aftermath of the plagues that got codified as the "natural" state.

To give just one example, when Captain George Vancouver traveled up the west coast of North America in 1792, he found a series of abandoned communities—large clearings in the forest that were newly growing over with "smaller shrubs and plants." In their travels, he and his crew found the bones of humans "in great numbers." Some were ceremonially buried in canoes or, for the children, in baskets hung in trees. But many bones were scattered, as if the people had died so quickly, they could not be laid to rest. The few living people they saw bore scars from what had killed the rest: smallpox. The disease had been carried through trading networks all the way overland from the east coast, where it was epidemic in 1775. It had reached the west coast by 1782, destroying whole cultures and disrupting countless ecological management regimes as it went.

Dismissing Indigenous land management as minimal, or treating pre-colonization ecosystem changes as "natural," is sometimes misperceived as a compliment to peoples who were able to husband their resources without depleting them (once you get past that first extinction wave back in the Pleistocene, that is). But the "virgin wilderness" narrative has been used around the world to deny Indigenous people rights to their lands. An applied anthropologist with the Tsimshian Nation, Brenda Guernsey, wrote in 2008 that "erasing First Nations landscapes and replacing them with a preconceived understanding of 'wilderness' allowed the landscape to be physically, socially, and conceptually cleared for the colonial settlement of the land." In Australia, British colonists took land that was being actively managed by Aboriginal peoples on the basis that it wasn't improved or culti-vated, and thus was *terra nullius*—no one's land. "Wilderness" thus isn't just a Romantic ideal, it's also a colonial power play.

Around the world, Indigenous people have been evicted from their homes, which were later rebranded as "wilderness" and set up as places where white people can recreate and relax. In California's Yosemite Valley in 1851, a unit of the California State Militia expelled a band of Ahwahneechee people, killing 23 and setting their houses and acorn stores on fire, to make way for gold miners. In 1864, president Abraham Lincoln made Yosemite Valley into a park—an act considered by many to be the beginning of the National Parks system.

Just four years later, naturalist John Muir came to California and fell in love with the landscape. He was less enamored of the Native people, whom he met while working as a shepherd in the high Sierra. He called them "dirty" and contrasted them with the "clean" wilderness. "They seemed to have no right place in the landscape," he wrote. He called Yosemite "pure wildness" and wrote that "no mark of man is visible upon it." Muir unsur-prisingly supported evicting yet more Native people when Yosemite was expanded and made a National Park in 1890.

What Muir didn't realize—or allow himself to understand—was that the landscape he loved so ardently was created by the "dirty" Indians he sneered at. As ethnobotanist M. Kat Anderson writes, "Staring in awe at the lengthy vistas of his beloved Yosemite Valley, or the extensive beds of golden and purple flowers in the Central Valley, Muir was eyeing what were really the fertile seed, bulb and green gathering grounds of the Miwok and Yokuts Indians, kept open and productive by centuries of carefully planned indigenous burning, harvesting and seed scattering." That these places were managed was apparently totally lost on him. Native people, he wrote, "hurt the landscape hardly more than the birds and squirrels."

Native people were pushed out of Yosemite. With the meadow keepers gone, trees encroached and the "parklike" open grounds turned bushy and cluttered with trees. In 1927, in the middle of the valley, they built the Ahwahnee Hotel, where today rooms start at $341 a night. In 1929, the last survivor of the 1851 massacre, To-tu-ya (Foaming Water), came back to Yosemite. In her 80s, she gazed over the valley floor. "Too dirty; too much bushy," she said.

————

Our concepts of "nature" and "wilderness" sadly limit the solutions that we can imagine. Perhaps because of the bluntly extractive tendencies of their ancestors, it remains very difficult for many people with primarily European ancestry to wrap their minds around even the *idea* of a positive, mutually beneficial relationship with other species. Thus they can see only two conceptual options: destruction of nature by humans or separation of humans from nature. To save nature, we must exile ourselves from it—like latter-day Adams and Eves leaving Eden in shame after despoiling it.

When Robin Wall Kimmerer asked her students to rate their own knowledge of "positive interactions between people and land," most told her they knew of none. "They could not even imagine what beneficial relations between their species and others might look like."

Rejecting the "nature" and "wilderness" frameworks does not mean rejecting the notion that some places and some individual animal lives are less influenced by humans, and others more. It makes sense to differentiate those places intentionally shaped by humans for human use from those places shaped by many species (including humans) and *not* designed for human use (but sometimes including some human use). If I could recast "nature" to mean something like this, I would. But when I say "nature" or "wilderness" people envision something *defined* by the lack of human influence, not by the presence of significant non-human influence. So I try to stick to words like "undeveloped" or simply "the outdoors" that have less purist connotations. Philosopher Ronald Sandler uses the phrase "non-built environments," which feels accurate.

Many people value "non-built environments" in part *because* they are less influenced by humans. I, too, delight in the "spontaneity and otherness" of these places, as Sandler puts it. But we can revel in the non-human without disdaining the human. And we can also find beauty and delight in relationships between humans and non-humans, from ancient harvest of sweetgrass, to cattle egrets following tractors to nab the grasshoppers they kick up, to suburbanites saving the eastern purple martin from extinction by putting up birdhouses for them.

This entangled world of mutual flourishing is what I meant by the "Rambunctious Garden" of my previous book's title. To make good environmental decisions, we must stop focusing on trying to remove or undo human influence, on turning back time or freezing the non-human world in amber. We must instead acknowledge the extent to which we have influenced our current world and take some responsibility for its future trajectory. Given that we actively use at least half the Earth's land for our own ends and actively manage many of our protected areas, the gardening metaphor seemed right. But I suggested that our global garden is and should be "rambunctious" because we must always leave room for the autonomy of non-humans. We should not seek to carefully control every plant and animal on the planet. We couldn't even if we wanted to.

Rejecting the human/nature dichotomy also does not mean condoning all human actions because as animals, everything we do is "natural." Using "natural" as a substitute for "good" is the problem here, not the solution. Many human actions have been bad for us, bad for other species, bad all around. As a collective, we humans have clearly taken more than our fair share of space, water, and other resources. But we don't fix that by exiling ourselves from the rest of Earth's species and building a wall between us. We fix that through repairing the systems by which we make our living, by learning—or re-learning—better, positive relationships with the species with which we share Earth.

Though I believe "wildness" and "nature" are incoherent concepts, I also maintain that as these terms are typically used, they do include two particular commitments that are worth retaining:

1. The flourishing of living things, which includes their autonomy
2. Human humility and restraint

To love the "wild" is to love the non-human in all its many millions of forms, to love the ways that plants and animals live, the choices they make, the beautiful patterns they weave as they exert their millions of individual agencies. To love "nature" is to love landscapes that remind us of our place as just one of millions of species on the planet. To call for the preservation of "the wild" in this sense is to call for humanity to think twice before wading into such complexities, full of cocky self-assurance, and making big changes. It is to argue that human needs and desires should not swamp those of other species. That is different from saying humans must be excised like a cancerous growth from any ecosystem in which our traces can be discerned.

I think we would actually be better at promoting the flourishing of autonomous living things with humility without the nearly religious worship of nature and wilderness. In some places, we value "naturalness" so highly that we become willing to hurt and kill animals to protect it. When non-native animals are killed simply because they "don't belong" *and not because they are clearly causing some measurable harm*, we have decided that erasing the taint of the human is more important than the lives of animals (who, lest we forget, have no conception that they are in the "wrong" place). This does not feel like humility in action. It is often the case that we hurt and kill animals because they are having effects we don't like, perhaps by predating on rare animals or eating rare plants. That's a trickier question—one we will tackle in due course. But I feel confident in saying that when we kill animals for no reason other than the fact that their presence isn't "natural" in that place, we are neither respecting the autonomy of living things nor showing humility and restraint.

––––––––

The value that humans place upon "wildness" and "nature" has hurt and restricted the autonomy of many individual animals—which brings us back to Thelma and Louise. What would have happened if humans had set aside their anxiety over the blurring of the line between "nature" and humans? What if the owners of the big sheepdog and the state of Washington simply let the dog answer his own *Call of the Wild* and form a pack with Thelma and Louise?

Certainly, there were some real pragmatic concerns that motivated the intervention in their lives. Their habituation to humans put them at risk of being shot—or of being run over by a car the way Thelma was. And if the mixed family had begun to hunt sheep, it would have become a bureaucratic nightmare. Was it a feral dog that killed this sheep and should it now be taken to the pound or put down? Or was it a wolf that we should haze away from all things human to reminded it of its wild role?

These concerns are valid. But what if the hybrid family learned to stay away from people? I like to imagine them just existing out there in the rolling mountains of the Colville National Forest, breasting snow drifts in pursuit of elk, denning in spring and playing with new pups each summer, free to make their own choices, flop-eared and wag-tailed, but socialized by their mother in the ways of wolfkind. They wouldn't be pure; they wouldn't be untouched by humanity; but I think they would have been wild.

5

The Lion in the Backyard

E lsa was a beautiful, well-muscled lioness, blocky of head, big-eared, and spotted on her legs and tummy. She was gentle, affectionate, and curious—a lioness who lived in both human and non-human worlds. Her story was told in 1960's *Born Free*, arguably the master narrative for positive human interactions with wild animals in the 20th century. The book was written by Joy Adamson and recounts the true story of how she and her husband made a lioness their pet and friend, and then successfully returned her to the wild. The story was also made into a film in 1966.

George Adamson, Joy's husband, was a game warden in Kenya, tasked with managing wildlife—and occasionally killing dangerous or nuisance lions. The Adamsons were white Europeans and the colonial system in which they lived is the unquestioned background for both the book and the film. On one of these expeditions, George brought back three tiny lion cubs he had just inadvertently orphaned by shooting their mother. In the book, there's a fantastic picture of Joy holding all three cubs in her arms with a look of absolute rapture on her face.

Two of the three cubs ended up in a zoo, but Joy couldn't bear to part with the youngest, Elsa. The young lioness lived with the Adamsons as a pet and seemed to have a pretty good life, riding on the roof of the Land Rover, exploring the landscape around their home, chasing elephants and giraffes, lounging during the heat of the afternoon under a tree while one of the Adamsons' servants, Nuru, read the Koran and kept an eye on her, before coming home in the evening to pre-killed goat or sheep meat. Joy does mention training Elsa not to jump on them with "the judicious use of a small stick," which they "seldom" had to use, though they "always carried it as a reminder." (The stick does not make an appearance in the film.) But Elsa seemed genuinely fond of them and Joy and George clearly related to Elsa as an individual, not just a generic example of her species.

They took Elsa with them on various trips around Kenya, introducing her to swimming in the Indian Ocean and ascending Mount Kulal, and the bond between the humans and the lion only grew. "To feel that we were responsible for such a proud, intelligent animal, who had no other living creature to satisfy her strongly developed need for affection and her gregarious instincts, attached us all the more deeply to her," Joy wrote. Joy added that, because they couldn't leave the lioness with anyone else, "we became to some extent her prisoners."

By the time she was 27 months old, Elsa was spending some nights away from home and the Adamsons were not quite sure what to do with her during their upcoming overseas leave, which they planned to spend in Europe. They decided to try to reintroduce Elsa to the wild. It was a lot of work. They had to travel to a suitable area, far from all human settlements, and train Elsa to hunt. They left her alone for longer and longer periods and brought her into close contact with wild lions to see if she could become integrated into a pride. Their first attempt failed, and they tried again in a new area. Eventually, they pulled it off. Elsa learned to hunt, began interacting with wild lions, and spent less and less time in their camp. One day they said goodbye and left, thereafter only visiting her new home for short stints. Elsa went on to have cubs of her own, which she brought to meet her human foster parents.

Joy Adamson's book was translated into 25 languages and sat on the *New York Times* nonfiction bestseller list for 48 weeks. It sold 6 million copies in its first year. The film was also an enormous hit. The actors who played the Adamsons, real-life couple Virginia McKenna and Bill Travers, were so inspired by their roles that they both became committed animal advocates and founded a charity now known as the Born Free Foundation, which campaigns to end the captivity of wild animals in circuses, their use in trophy and canned hunting, and their exploitation in wildlife trafficking. "I don't think really, before [*Born Free*], animals were ever looked at as individual beings," McKenna said in a 2010 interview.

Among the first activities of the Born Free Foundation was securing funding for George Adamson to rehabilitate the lion actors in the film, with a view towards reintroducing them to the wild as well. More captive lions found their way to him over time, and he ran a kind of free-range sanctuary for many years. When his favorite, "Boy," killed a staffer, George had to shoot him. "As I have learned at great cost, it might be true to say that no lion is completely reliable. But are many human beings either?" he wrote. Ultimately, Joy and George Adamson lived apart, and both were killed by humans—Joy in 1980 by Paul Nakware Ekai, an employee who accused her of shooting him in the foot for borrowing her land cruiser without permission and George in 1989 by bandits near his lion camp.

Today, one of the Born Free Foundation's central fundraising efforts is to ask the public to "adopt" a wild animal for a monthly donation. Instead of getting a live lion lounging on your sofa, though, you'll receive "an exclusive adoption pack" containing a soft toy, an animal story, species fact sheet, personalized certificate, and a glossy photo. This approach taps into the human desire to have a good relationship of some kind with a wild animal without the considerable downside. The money goes to a sanctuary for rescued captive lions in South Africa and conservation projects in Kenya.

The *Born Free* story is indeed remarkable and inspiring. I, like so many before me, burst into helpless tears at the end of the film when Elsa appears with three adorable cubs. But the expectations that it sets up are problematic, because Elsa's story is not typical. In fact, it's almost magically untypical. Very, very few pet big cats have been successfully reintroduced into the wild from captivity. Even conservation organizations with big budgets and scientific boards have struggled to do what the Adamsons did. You will notice, for example, that the Born Free Foundation is running a lion sanctuary, not attempting to reintroduce the animals they rescue to the wild. Lions, in particular, don't rewild easily.

The Adamsons made it work because Elsa had grown up exploring the ecosystem into which she would eventually be released. They made it work because they were professional wildlife managers and because they labored incredibly hard at it for months and months. Elsa's individual personality, including her high levels of curiosity and tolerance for novelty, probably helped too. The Adamsons had it both ways. They got to live with a lioness, to be her family, and know her intimately—and they were also able to give her a full, wild life. It usually doesn't work this way.

Thus when investigative journalist Rachel Nuwer interviewed a woman named Deborah Pierce in South Carolina with a full-grown lioness penned up in her backyard, I was grimly unsurprised to hear that Pierce had named her pet "Elsa." The irresistible promise of *Born Free* is that you can be pals with a big, beautiful wild animal without permanently harming it. Pierce now regrets buying the lion, in part because of the thousands of dollars of meat and veterinary care it needs every year and in part because conflict over the lion played a part in her divorce. She is paying the price for falling for *Born Free*'s magic spell without fully absorbing how unique the case was. And the lion behind the chain-link fence in her backyard is paying the price too.

The promise of *Born Free* is that there is a way to have a good relationship with wild animals that is interactive, that is more than simply leaving them alone. It suggests that we can *have* a wild animal. For those who truly love wild animals, it is understandable that there's a desire to be close to them, to become friends, even family, with them. And if all "wild" animals are already highly influenced by humanity, then perhaps there's less of a reason to stay strictly hands off. But just because we're not buying into a bright line between humans and all other species doesn't mean that all interactions with animals are ethically permissible. On the contrary, respecting the autonomy of individual animals instead of focusing on the purity of their "wild" pedigree suggests that any positive relationship between us and them must be by mutual consent.

Pierce told Nuwer that she bought the lion off the Internet for $1,500. It had aged out of a cub petting operation. If you lived in the United States in the 1970s or 80s, you might remember a phenomenon called "zoo babies." As Nuwer writes, "Each spring, interstate signs and TV commercials featured photos of blue-eyed, squealing balls of fuzz debuting at major zoos around the country, a powerful marketing lure for families that turned out to snap photos and cuddle the newest arrivals. Zoo babies were among the industry's number one moneymaking programs, and tigers were always the biggest draw." But the zoos didn't have room for all the babies once they grew up. All those cuddly cubs became "zoo surplus." And zoos happily sold them to private parties, fueling an explosion in private ownership of big cats.

Many people who bought such cats did so because of a sincere belief that they were somehow helping conserve the species—although no big cat conservation project currently underway would even consider using formerly captive tigers or lions with mixed-up genetics—called "generics"—in their work. Like wildlife managers struggling to keep wolves and dogs apart, they would want animals with pure bloodlines—all Siberian tiger or all Bengal tiger, for example. Even if conservation organizations weren't hung up on genetic purity, reintroducing captive animals to the wild is extremely difficult, as we've seen. Most conservationists would much rather see existing wild populations grow.

Other customers for the big cubs, like Pierce, just adored their beauty and their power, and wanted to take that "wild" majesty home with them. Many ended up in the hands of showmen and -women, who breed them and use them to profit off our deep-seated need to touch the wild. These "roadside attractions" are in a bit of a gray area between "zoos" and "pet" ownership. Such operations increasingly use words like "conservation" or "sanctuary" in their names, but they are really no different from a circus sideshow: People pay to see and touch the animals. It's a straightforward business transaction. Mix our inborn urge to care for the small and vulnerable with our desire to take risks and seek novelty, add a dash of craving for a positive connection with non-human animals, and—for the garnish—offer a social-media-worthy photo op, and you have yourself a potent cocktail.

Big cat cubs are only safe to handle for a few months, so breeders continuously produce more, weaning the cubs while their eyes are still closed so their mothers can breed again. The cubs are then put to work as photo props and cuddle objects for up to 10 hours a day, according to an investigation by Sharon Guynup, a Global Fellow at the Wilson Center. Once they are too old to be handled, a few graduate to become breeders or display animals, but many just disappear, likely killed. Guynup found that these operations can make over a million dollars a year. Today, there are an estimated 3,900 tigers in the wild, some 7,000 in "tiger farms" in Asia—and up to 10,000 in captivity in the United States.

This is the basic business model behind outfits like the G.W. Exotic Animal Park in Oklahoma, the demesne of the "Tiger King," subject of the wildly popular Netflix docu-series that entertained so many during the early days of the coronavirus stay-at-home orders in 2020. Joseph Maldonado Passage, aka Joe Exotic, acquired his first tigers from another private owner who was keeping them in the backyard, along with a leopard and a mountain lion. By the time of the series, Joe was breeding them, and charging people to pet the cubs. If he had ever been an animal lover he had long since become far more interested in his own personal brand—and the prospect of being the subject of a reality TV show. Eventually, he wound up in prison on a 22-year sentence for trying to hire someone to kill Carole Baskin, who runs a legitimate sanctuary for big cats and was trying to shut him down, as well as for illegally killing and transporting big cats across state lines. Thousands who tuned in to the series saw Joe as a kind of folk hero, a gay redneck polygamist who just happened to own dozens of tigers. But his animals led sad, small lives to support his narcissistic life. He exploited both his employees and his animals ruthlessly.

———————

The urge to co-habitate with wild animals seems to cut across cultures. When Europeans first met Native Americans, they noted that they kept a wide variety of pets, including "raccoons, monkeys, peccaries, tapirs, wolves, bears, moose, mice, rats, squirrels, and birds." Aboriginal peoples in Australia have long kept "wallabies, possums, dingoes, bandicoots, and cassowaries," and Indigenous peoples in South America keep "agoutis, pacas, parrots, rodents, sloths, and monkeys." Across the world, animals brought home and kept alive become members of the household.

An influential 1994 paper by Elizabeth Hirschman attempts to detangle the many motivations that humans have for bringing non-humans into their homes. Broadly, she says, pets can be divided into "animals as objects" and "animals as companions."

In the first category, she includes animals kept as "ornaments" for their sheer beauty. Think about the tank of colorful fish or birds kept for their beautiful plumage or songs. Another category of "objects" are those animals kept as status symbols. In the Gulf states, rich men post Instagram photos of themselves with their pet cheetahs, illegally smuggled from a dwindling wild population. Gayle Burgess, senior coordinator for behavior change at TRAFFIC, a nonprofit that fights illegal and unsustainable trade in wild animals and plants, told me that in Indonesia, songbird ownership is about your status as a man. "There are fierce competitions between men based on the beauty of a bird's song," she says.

Pop star Justin Bieber has two savannah cats named Sushi and Tuna. These cats are a hybrid between servals and domestic cats. They have their own Instagram account. Previously, the singer owned a capuchin monkey named OG Mally, but it was seized in 2013 by German authorities because it didn't have proper documentation and it looks like Bieber never went back for it. "People are always like, 'Why did you get a *monkey*?'" Bieber was quoted as saying. "If you could get a monkey, well, you would get a fucking monkey, too! Monkeys are awesome."

Nuwer, whose excellent 2018 book *Poached* covers the international wildlife trade, says that many people who buy wild animal products are primarily interested in the prestige that comes with owning a high-dollar and difficult-to-obtain luxury good. Pulling out some rhino horn powder at a party can enhance your status like a Patek Philippe watch or Birkin bag. Thus as anti-poaching efforts and stricter anti-trafficking laws make animal goods harder to get ahold of, they can also make the animals and the products derived from them more valuable. Precisely *because* wild animals are increasingly rare, they are extravagant symbols of the wealth and power of the owners.

Then there are animals who are valued by humans for their companionship. Hirschman says the tenor of this bond varies considerably, from the pet who is seen as a friend and partner to the pet treated as a sort of child. Hirschman delineates a further type within the general "companionship" category. She says some animals "act as extensions of the consumer's self." Here, "the animal's traits, behaviors, and appearance are seen as being those of its owner; the owner projects his or her own personality onto the animal and absorbs the animal's nature into himself or herself."

Many exotic pets in postindustrial Western countries fall into this category, at least in part. The animal is both a pal and an extension of the personal brand. The fact that the animal is "wild" speaks to something inside the person, perhaps a sense that they too are wild, free, powerful—or that they wish they were. When pop musician DJ Khaled posed with a lion for the cover of his 2016 album *Major Key*, he said the lion was a symbol of himself, of "what I represent, from my spiritual vibes to my beliefs. And that is a certain energy that I keep around me: positivity. But at the same time, as a king. You know what I'm saying."

Lisa Wathne, who studies captive wildlife at the Humane Society, says the exotic pet owners she sees tend to fall into the "status symbol" and "companionship" categories. "With tigers and bears, you are dealing with mostly men who like to give the impression that they are able to control or handle this dangerous animal," she told me. "It's an ego thing." But some also see their wild pets as part of the family. "With the primates, it is a different thing," Wathne says. "We call them the monkey moms. They become surrogate children. But once they outgrow that baby stage and they reach puberty, they get aggressive and all too often end up in tiny cages. It is tragic what happens to these animals."

Some people love animals of a particular species so much that they seem unable to help themselves, even if they know the rules and risks. They simply must have them. The decision might be split-second, with people finding animals for sale and being overcome with the desire to possess them—or even to "save" them, according to Burgess. Imagine strolling through a market on a hot day and seeing a monkey in a little cage, looking sad and weak. "To some extent maybe you want to rescue the animal because it looks heat stressed," she says. "A lot of people really genuinely love animals and want to be close to them," Nuwer told me. "The idea of being close to the wild and tapping into our natural selves is really compelling. It is trying to fulfill some vague longing that some of us have inside of us."

———

The world's most popular pet is the dog, so it makes sense that some people want the wilder version, keeping wolves or wolf-dog mixes as pets. At first glance, this seems potentially less cruel than dressing monkeys up in baby clothes or penning big cats up in the backyard. After all, dogs were once wolves, so the ability to bond with humans and thrive in a human world seems like it could be latent in all *Canis lupus*. If a sheepherding dog could vault its fence and "be a wolf for a month," maybe a wolf can make the jump in the opposite direction and become a happy part of a human household.

The history of interbreeding between wolves and dogs stretches back to the origin of dogs, since the domestic dog derives from the wolf, as many as 40,000 years ago. Throughout history, dogs and wolves have met and bred—sometimes on their own, like Thelma and the sheepdog, and sometimes at the will of humans, as when the ancient Gauls tied their female dogs to trees to make them available to local male wolves. Humans continue to breed wolves with dogs intentionally, and in the United States, wolf dogs are available for sale in the states where they are legal (and in the states where they are illegal).

According to an article on wolf dogs as pets in the doglover's magazine *The Bark*, the appeal of the wolf dog is the idea of "a dog's friendly companionship paired with a wolf's good looks and untamed nature. Buy a wolf dog, the thinking goes, and live out your Jack London fantasies, even if you're in Akron rather than Anchorage." The piece goes on to quote the author of a guide to owning wolf dogs on the motivations of her readers: "They want to own a piece of the wild, and they often say that the wolf is their spiritual sign or totem animal." The paradox of "owning" part of the wild must either appeal to such people or not occur to them.

Wolf dogs are much frowned upon in the dog world, however, chiefly because it is felt that their temperaments are unsuited for life as pets, creating a miserable situation for both animal and owner. Even intensive, thoughtful socialization cannot overcome genetically coded traits like low thresholds for stress hormone release and, perhaps, fear and instinctive avoidance of humans.

Ceiridwen Terrill had just ended an abusive relationship with a man when she began a new relationship with Inyo, a wolf dog she raised from a pup. Why a wolf dog? In part, she liked the idea of having a dangerous protector from her violent ex. But it was more than that. Terrill wanted a wolf dog not just as an extension of herself but as an aspirational self: unafraid, fierce, free.

Inyo was mostly wolf and a little husky. She was narrow-bodied, with a thick brushy tail that hung straight down, a white face, and dark amber eyes. (All wolves have amber eyes. When you see a "wolf" with blue eyes as a pet or in a movie or commercial, it is probably a husky.)

Ceiridwen adored her, but their life together was difficult, as she explains in her 2011 memoir, *Part Wild*. She describes wolf dogs as a *zweiweltenkind*—in German, a child of two worlds. Because Inyo was raised by humans rather than wolves, she would have been helpless if released into the wild. But she was unhappy at home too, relentlessly peeling back the linoleum and trying to dig into the wooden floor, eating electrical cords, and keeping neighbors awake with her howls. She constantly tried (and often succeeded) to escape from her kennel and yard—likely driven by the biological instinct to disperse, find a mate, and carve out a territory, but she had nowhere to go. She'd wander the neighborhood, getting into trouble.

Eventually, Inyo began to bite dogs and people. Terrill called sanctuaries, looking for a place that could take her pet. One day, Inyo bit Terrill on the ear and arm. "I pressed a dish towel to my bloody ear, feeling sick," she writes. The attack was not really Inyo's fault. Ultimately, it was the fault of those who had bred her. "Inyo's birth had been the result of a terrible human error." After one more hike together through the mountains, one more swim in the creek, Terrill took her to the vet to be euthanized. Inyo, Terrill writes, "taught me to appreciate wildness and to leave it alone."

Most people simply do not have the space, the money, or the knowledge to provide a wild animal with everything that would make them happy, even if they love them dearly. A key issue for larger species is space. Backyard big cats and wolf dogs live in much, much smaller spaces than their wild relatives. Wild tigers maintain territories from three to nearly 400 square miles. Wolf families maintain home ranges of up to 2,450 square miles, depending on the size of the family, the density of prey, the fierceness of the competition, and other factors.

Captive tigers notoriously pace back and forth, and researchers have found that "the time devoted to pacing by a species in captivity is best predicted by the daily distances traveled in nature by the wild specimens." It is almost as if they feel driven to patrol their territory, to hunt, to move, to walk a certain number of steps, like they have a Fitbit in their brains.

Exotic pets can be dangerous. The Humane Society has compiled a list of two dozen people killed by big cats since 1990, but there is no systematic collection of data about wolves or wolf dogs. In the absence of good data, a 2006 story from the *Pittsburgh Post-Gazette* haunts me. A 50-year-old woman was found dead "in a caged rectangular enclosure in her backyard, home for nine half-wolves, half-dogs." She had bled to death after the hybrids attacked her. The article went on to quote her neighbors, who said she treated the dogs "like children" and added that she had told them that the wolf dogs "give me unqualified love."

Of course many wolf dogs probably do love their owners, in their own way. But from what I can tell, a distinctive feature of wolves and many wolf dogs is that they *just aren't that into us*. Unlike dogs, who are predisposed to bond with humans and who constantly look us in the eye, searching for clues about what we want and how we are feeling, wolves and hybrids just don't care that much about us and may actually instinctively fear us.

Wolf Haven International, where Ione lives since being parted from the dog with which she wished to form a pack, is mostly populated by ex-pets. Executive director Diane Gallegos estimates that 80 percent of wolf dogs are dead by age two—euthanized because owners could not handle them and shelters would not take them. Gallegos adds that many wolf dog owners are young men who are interested in a "macho" pet or women who see wolves as their "spirit animal." (Gallegos is quoting the pet owners here. Indigenous people from tribes with actual spirit animal traditions have made it clear that they find it hurtful and offensive when people who aren't tribal members appropriate the phrase. Some alternatives to use instead are "personal icon," "familiar," "alter ego," or simply "favorite.")

When I asked Wolf Haven how many pet wolves and wolf dogs are out there, they referred me to the Humane Society. So I asked Lisa Wathne at the Humane Society, who suggested I ask Wolf Haven. When I told I already had, she threw up her hands. "With no one agency keeping track or no require-ments for people even to report they have these animals, in most cases, there's simply no way to know how many are out there," she said. "It's a mess."

Wanting to see these *Zweiweltenkind* in person, I decided to visit Wolf Haven, and I brought my husband and two small kids along. We walked by the enclosures of animals on the public route. London was a white wolf retired from the film industry. He loved to stalk visitors. Juno was a tall wolf dog who was sold as a "pure wolf" pet. When she got to be sexually mature, she ripped the bumper off a Lexus SUV. "They have been forced into a world where they don't fit," Gallegos told us. The wolf and the dog inside them can conflict, leading to psychological problems. The wolf in them is afraid of people; the dog is drawn to them. The wolf needs to strike off on its own; the dog is happy in a human home. "You can see the mental instability in their eyes," Gallegos said.

Many of the former pets at Wolf Haven are weaned from human contact and paired with other canines to fulfill their social needs. They are never bred. As we were chatting, one of the residents began to howl and the other wolves and wolf dogs began to join in. With enclosures all around us, we were surrounded by a polyphony of long mournful tones, each note bending at the end. We humans fell silent. My kids instinctively edged closer to my husband, and he crouched down to put his arms around them. The howl swelled around us, running up the back of my neck into my hair, then died away. Later, I asked my four-year-old what she thought when all the wolves howled. "I thought I was in the wild," she said. "I thought I was a wolf."

———

The staff at Wolf Haven do the best they can to make the canines living there happy. But they freely admit that the animals' lives are not ideal. There's a difference between happiness and true flourishing. In philosophical circles, the concept of "flourishing" is meant to describe what it means to live a good life. Most of the things we think of as "valuable," like money or good health, are really just tools for achieving flourishing. Flourishing goes beyond happiness to include such things as having a life filled with meaning. Aristotle believed that humans flourish when we exercise our capacities as humans. He reasoned that since we had unique powers of rational thought (or so he believed), using our rational minds to live a virtuous life was the height of human flourishing. Of course a philosopher would say that flourishing consists of thinking well!

But what is flourishing for an animal? Is it more than just happiness? Does animal flourishing include individual autonomy? I've argued that the autonomy of individual animals is valuable in a way that the purity of their "wild" genes is not. If I'm right, then controlling or keeping captive wild animals that could live autonomously may be wrong even if they are well cared for. For most wild "pets" it is too late for complete autonomy. Without a childhood spent learning the ways of their kind, they cannot live happily in the wild. A sanctuary like Wolf Haven is their best option. But by refusing to capture animals for the pet trade and refusing to breed those already in captivity, humans could eventually put sanctuaries out of business— presumably the end goal of every true sanctuary.

How can we know if non-human animals are flourishing? It is easy enough to look at a Labrador retriever asleep by the fire, its sausagey tail thumping faintly to the rhythm of its dreams, and feel confident that the animal is, at the very least, happy. Sitting in my living room, looking out at a flock of American robins gorging themselves on crabapples, it seems clear that they are having a wonderful time. They are making a furious racket, singing and gobbling and flapping from branch to branch. Some are so full that they are waddling around on the ground, looking sleepy. But how much of the joy that I see in their behavior is me projecting my own mind into their little skulls?

Philosopher Martha C. Nussbaum focuses on the "capabilities" of an individual to determine what flourishing means to them. Hers is a justice-focused approach, closer in spirit to Regan than Singer. If *rights* are the things that everyone is entitled to, *justice* is the practice of making sure everyone gets those things.

Nussbaum argues that *everyone* should get to exercise all their core capa-bilities, and no one should have their ability to do so trampled upon for the greater good. For humans, she identifies ten core capabilities, including having good health, feeling emotions, and controlling one's own environ-ment (a capability that can cover a range of things, from having secure title to one's land, to being able to vote). "A society that does not guarantee these to all its citizens, at some appropriate threshold level, falls short of being a fully just society, whatever its level of opulence," Nussbaum writes.

A sentient animal's core capabilities might look quite different. It depends on the animal, Nussbaum says. But she does suggest a few ideas, developed through the admittedly imperfect technique of "sympathetic imagining." First she imagines the lot of a circus animal, "squeezed into cramped and filthy cages, starved, terrorized, and beaten, given only the minimal care that would make them presentable in the ring the following day." Then she imagines what they would want or need. "Dignified existence would seem at least to include the following: adequate opportunities for nutrition and physical activity; freedom from pain, squalor, and cruelty; freedom to act in ways that are char-acteristic of the species (rather than to be confined and, as here, made to perform silly and degrading stunts); freedom from fear and opportunities for rewarding interactions with other creatures of the same species, and of different species; a chance to enjoy the light and air in tranquility."

If justice requires that we allow animals these components of a "dignified existence," then preventing wolves and tigers from dispersing from home, meeting mates, finding wild prey and killing it, establishing and defending territories—all these are unjust and wrong. By preparing the original Elsa for wild life, then returning her to the bush, the Adamsons, at great cost to themselves, did the right thing by her. Once free, she was able to mate, be a mother, hunt, and express her lioness self—although she never joined a pride. When she later came at the sound of a gunshot to visit with the humans she grew up with, she did so of her own free will, demonstrating that *she* valued the relationship. At the end of her life, she came into George's tent, and died there, with her head in his lap.

Saying that animal flourishing is valuable doesn't mean it is the *only* valuable thing in the world. There may be cases when protecting or promoting other valuable things is important enough that it may justify some infringements on the flourishing of animals. We'll take a look at some of those cases. But I don't think a convincing case can be made that the pleasure of owning a wild pet is justification enough.

I recognize in these misguided attempts to bring wild animals into our families a clear impulse to create good relationships with non-human animals. This impulse is an important clue about something that is missing in our culture's current framework for thinking about wild animals.

However, keeping a tiger or a wolf as a pet because we love their species or because we love "wildness" is a terrible idea. Many of these pet owners feel that they are being loving, but their love is missing the element of respect that makes it a virtue. Imposing the intensive human interaction that a dog would welcome on a wolf that wants nothing to do with us is cruel. I am convinced that private individuals, by and large, should not take wild animals home.

But what about scientific and conservation institutions who are trying to educate the public or save species? Elsa returned to the wild, but her siblings Big One and Lustica lived out their days at the Rotterdam Zoo. Joy wrote that "they live in splendid conditions." At least one of them was bred, and had at least seven cubs, all of which presumably spent their entire lives in zoos. Can modern zoos provide a kind of captivity that would allow animals to express their core capabilities and flourish?

6

The Autocratic Menagerie

Around Christmastime 2019, my father gave me some free passes to the Woodland Park Zoo in Seattle. I had become increasingly uncomfortable with zoos, so I hadn't been in a while. But hey, the tickets were free.

On their website, the Woodland Park Zoo says that their mission is to "save wildlife and inspire everyone to make conservation a priority in their lives." Like most modern zoos, they emphasize the conservation aspect of their institution. The Woodland Park Zoo makes much of its endangered species, like their greater one-horned rhinoceroses and their unbelievably cute and tiny Egyptian tortoises. But they have plenty of common animals too. The ones that get to me, perhaps unsurprisingly, are the wolves. They have a "pack" of white Arctic wolves on display, who seem to endlessly patrol their tiny territory against incursions that will never come. Screwed onto the wooden railing surrounding their enclosure is a small brown sign: "Thank you for not howling at the wolves." As I photographed the sign, a family arrived. As soon as the kids saw the wolves, they instantly started howling. I wondered: Do we howl to connect with the wild or to demonstrate our power over these animals? And what does this ceaseless, jocular, human howling do to the wolves?

Controlling wild animals, especially very large herbivores and frightening carnivores, has long been a way to show strength. Capturing dangerous wild animals may have begun as a precursor to ritualistic controlled hunts, like the lion hunts undertaken by kings in Mesopotamia as far back as 3,000 BCE to demonstrate their might and ability to defend their people from the dangers of the wild. Ancient Egyptian rulers kept collections of wild animals, which they sometimes brought out onto the battlefield with them. Ramses IX sent live hippos, monkeys, and crocodiles to an Assyrian king as a slightly aggressive gift. The animals were not only symbols of power, but of the whole non-human world. To capture and control them was thus to make a statement about who was in charge. As philosopher Stephen R. L. Clark writes of the canned hunts put on in the Colosseum in Rome during the Empire, "wild beasts were 'hunted' and killed to prove, dramatically, that human beings, and particularly *Roman* human beings, control the world."

Later, captive animals turned up at the Tower of London. Power-mad men from Henry III to Saddam Hussein's son Uday to drug kingpin Pablo Escobar to Emperor Charlemagne have all tried to show their strength by keeping terrifying beasts captive.

It is these boastful collections of animals, these autocratic menageries, from whence the modern zoo, with its didactic plaques and $15.00 hot dogs, springs. The first modern zoo was probably the London Zoo in Regent's Park, which opened in 1828. At first, it was only open to fellows of the Zoological Society of London, but it quickly expanded its customer base and provided novelty and amusement to the burgeoning Victorian middle class, serving up the fauna of colonized lands for a quasi-educational after-noon out. Public zoos multiplied and became points of civic pride, reflecting the prestige of a city the same way a captive lion or two used to symbolize the might of kings and empires.

After the London Zoo opened, similar public zoos sprang up all across Europe. They were ostensibly places for genteel amusement and edification and they expanded beyond the big and fearsome animals to include reptile houses, aviaries, and insectariums. Living collections were often presented in taxonomic order, with various species of the same family grouped together, for comparative study. Architecturally, European zoos leaned into the exoticism of wild animals from abroad, adorning their buildings with minarets and pagodas and faux Egyptian temples.

Some zoos exhibited colonized and Indigenous people. Animal importer Carl Hagenbeck exhibited Sami people in traditional garb along with reindeer in Hamburg in 1875. "Our guests," Hagenbeck wrote, "were unadulterated people of nature." He followed up with tours of European zoos showing people from Sudan, Greenland, and Sri Lanka. "Exotic" foreigners had been exhibited in Europe many times over the centuries, but here people were displayed, often alongside animals, in zoological parks. Hagenbeck insisted his shows were scientific and that his people weren't performing; they were just existing in their natural state.

In the most notorious case of humans in zoos, a Mbuti man named Ota Benga who was kidnapped from his home in the Congo Basin and sold into slavery was exhibited at the Bronx Zoo's Monkey House in 1906. He was freed after a public outcry and moved to Virginia, where he worked in a tobacco factory and hunted alone in the woods. But he struggled with the intense trauma he had endured and in 1916 he shot himself in the heart.

The same Carl Hagenbeck who exhibited people like wild animals also changed the way wild animals were exhibited. In his Animal Park, which opened in 1907, he designed cages that didn't look like cages, using moats and artfully arranged rock walls to invisibly pen animals. By laying these enclosures out so many animals could be seen at once, without any bars or walls in the visitors' lines of sight, he created an immersive panorama, in which the fact of captivity was supplanted by the illusion of being in nature oneself. Historian Nigel Rothfels says Hagenbeck's human and animal exhibits were both produced from the same worldview. "Visiting Hagenbeck's animal exhibits and people shows, visitors were not confronted with scathing critiques of capitalism, imperialism, or colonial exploitation—this was a idealized world where Europeans could walk among the exhibited animals and people and feel comfortable, secure, and, of course, enlightened."

Hagenbeck's model was widely influential. Increasingly, animals were presented with the distasteful fact of their physical control visually elided. Zoos shifted just slightly from overt demonstrations of mastery over beasts to a narrative of benevolent protection. From here, it was an easy leap to focusing on protecting animal species.

The "educational day out" model of zoos endured until the late 20th century, when zoos began actively rebranding themselves as serious contributors to conservation. Zoo animals, this new narrative goes, function as backup populations for wild animals under threat, as well as "ambassadors" for their species, teaching humans and motivating them to care about wildlife. This conservation focus is now *mandatory* for institutions that want to be accredited by the Association of Zoos and Aquariums (AZA), a nonprofit association of zoos and aquariums in the United States and 11 other countries.

This rebranding has mostly been applauded, although the new "naturalistic" enclosures meant that sometimes it was hard to see the animals, to the consternation of the public. The visitors' desire for the animals to be comfortable was not always greater than their hunger for intimate interaction—for touching, response, eye contact. I believe that it is this deep desire that drives some zoo-goers to throw things at the animals. People are desperate for connection with other animals, even if through the currency of pain.

Animal rights theorists like Tom Regan are opposed to zoos, in general. If we must respect animals as "subjects-of-a-life" and not treat them as means to an end, then requiring that a tiger or an elephant spend its entire life in a cage so it can be an "ambassador" for its wild cousins seems unthinkable. Keeping animals in zoos could only be justified if it is in *their* best interests. Regan himself sketches out a hypothetical case where an animal is taken into a kind of protective custody until a threat of "human predation" is eliminated. But these situations would be temporary and acknowledged as less than ideal. *Breeding* non-endangered animals just for display would be forbidden.

It you are a utilitarian like Peter Singer, then much rests with whether animals would prefer to stay in zoos or leave—as well as how much pleasure they give to how many people. Some individual zoo animals might have a pretty good, safe life. Those species that don't need a lot of space and that simply want a bit of habitat and plenty of tasty food might even sign up for it, if they could rationally weigh their options. A ground squirrel, for example, rarely strays further than 250 feet from its burrow, and everybody—from snakes to coyotes to hawks—loves to eat them, so they might theoretically prefer a large naturalistic zoo enclosure with plenty of food and places to dig. But people don't really go to zoos to see little animals like ground squirrels, do they? They want to see big impressive animals: tigers, lions, polar bears.

Former Woodland Park Zoo director David Hancocks wrote in 2001 that zoos, "expose a perpetual dichotomy, which is the reverence that humans hold for nature while simultaneously seeking to dominate it and smother its wildness. They reveal both the best and the worst of human nature."

The 2013 documentary *Blackfish* opened a conversation about whether captivity is acceptable for killer whales, which are both physically huge and cognitively complex. *Blackfish* focused on the 12,000-pound Tilikum, a massive animal with the telltale collapsed dorsal fin of the captive male orca. After being taken from his family in Icelandic waters at age two in 1983, Tilikum spent the rest of his days in glorified swimming pools, performing for humans, and siring 21 children destined for a similar life. He was involved in the deaths of three people before dying in 2017—and he wasn't the only captive orca to attack or intentionally drown humans.

The film was so convincing in its portrait of a prisoner driven mad by his imprisonment that the general public seems to have turned decisively against keeping orcas in aquariums. SeaWorld saw attendance crater. The water park knew when it was beaten. They announced they would stop breeding orcas in 2016. Since the United States hasn't issued a permit for the capture of a wild orca since 1989, this suggests that by around 2050, when their current orcas have all died, there should be no more captive killer whales in the franchise. Joel Manby, then SeaWorld CEO, suggested that the parks would pivot to working with animals that were injured or orphaned in the wild. "We will increase our focus on rescue operations—so that the thousands of stranded marine mammals like dolphins and sea lions that cannot be released back to the wild will have a place to go," he said in a statement.

SeaWorld is still making money on the 20 orcas they have left. Today, you can "connect in an inspiring new way with the ocean's most powerful predator" at the "Orca Encounter" show at SeaWorld in Orlando, included in the $54 ticket price. After all, as their website says, "Killer whales are the perfect ambassadors for the ocean."

If the Whale Sanctuary Project has its way, SeaWorld's orcas—and the 3,000 other whales and dolphins in captivity around the world—won't live out the rest of their lives in tanks. In 2020, the nonprofit selected a site in Port Hilford, Nova Scotia, where they plan to net off a bay or cove in the ocean for cetaceans that can't live independently in the wild.

While the average orca tank at a SeaWorld-type park is around 10,000 square feet, the outdoor ocean enclosure would be at least 2.8 million square feet. And the project is more complex than just stringing up some netting. The charity is working to secure permits to host marine mammals at the site and fundraising to pay for an "animal care center with full-time veterinary services, freezers for the fish the whales will eat, offices for researchers and staff and security personnel, as well as the kinds of behind the scenes planning that is so necessary for water service, electricity and road access." They estimate it will cost "$12-15 million U.S. for the creation of the sanctuary, and then $2 million U.S. per year for the care for 6-8 whales." In comparison, SeaWorld's revenue in 2019 was 1.4 billion dollars.

Elephants are the largest land mammals, and they are known for their intelligence—especially their emotional intelligence. It makes sense that concerns about animal captivity would begin with these majestic animals. The Woodland Park Zoo featured elephants on and off as far back as the 1920s, when the schoolchildren of Seattle donated their pennies to help buy Cleopatra, a petite Asian elephant from a vaudeville act called "Singer's Midgets."

Over the years, the Woodland Park Zoo has had a dozen elephants, including Tusko, who was rescued from very poor treatment in a traveling circus, and Elmer the Safety Elephant (actually a female) who was supposed to convince children to cross streets at crosswalks, and who died of a tusk infection at age 17. But the elephant era began to end with the 2012 publication in the *Seattle Times* of "Glamour Beasts: The Dark Side of Elephant Captivity" by Michael J. Berens.

The investigation focused on Chai, an Asian elephant who was captured before she was weaned and given to the city of Seattle in 1980 by Thai Airways to celebrate their purchase of a Seattle-built Boeing 747 airplane. The zoo director, David Hancocks, didn't think elephants really belonged in zoos. He was already caring for Bamboo, captured in Thailand as a calf in 1967, and Watoto, who had come as a baby from Kenya in 1971. He wanted to reject the gift, but he was overruled by the mayor. Concerned that the elephants would be bored cooped up in their "barn," a "drafty, leaky, uninsulated building," given that wild elephants walk up to 50 miles a day, Hancocks let them stroll around the zoo and the larger Woodland Park with their keepers. "I recall one summer afternoon when there was concern because Bamboo and her keeper had not been seen for some time," Hancocks reminisced in the *Seattle Times*. "We found them lying in a sunny glade, down by Aurora Avenue North, each fast asleep, the keeper propped against her comfortable girth."

When Hancocks couldn't get as much space as he wanted for a new elephant habitat, he resigned in frustration. Afterward, he became a harsh critic of how the elephants were handled. Under his replacement, they were chained up at night and physically "disciplined." There were no more leisurely strolls.

In the 1990s, the zoo decided to use artificial insemination to try to impregnate Chai, chaining her in place and winding a tube full of sperm up into her body as often as ten times a month. It didn't work. They then sent her to the Dickerson Park Zoo in Missouri to mate with a bull elephant, despite knowing that elephants there had a herpes virus that could infect Chai and threaten any baby she might have. They decided that the prospect of a baby elephant was worth the risk.

While in Missouri, Chai was chained up and "disciplined" with ax handles and a bullhook, a long stick with a hook at the end. The zoo even paid a fine to the United States Department of Agriculture for violating the Animal Welfare Act by causing Chai "trauma, behavioral stress, physical harm and unnecessary discomfort." Other elephants fought with Chai.

Keepers dosed her with Valium and azaperone to keep her calm and tractable. Although Chai lost 1,300 pounds in Missouri, she returned to Seattle pregnant. The zoo was thrilled. Her baby, Hansa, was born in 2000, weighing 235 pounds, and looking undeniably adorable.

Tiny Hansa was a huge draw for the zoo. She wasn't quite tall enough to reach Chai's nipples, so the zoo made her a nursing platform. She was famous for playing with enormous rubber balls—almost frisky, for an elephant. In her first year of life, 1.2 million people came to see her. According to Seattle's alternative newspaper, *The Stranger*, zoo employees nicknamed her "Cash Cow." She was trained to come when called, back up, turn in a circle, raise her trunk, and hold up her feet to be inspected. According to David Hancocks, that training involved being hit with a bullhook when she was just a few months old. Later, at 18 months, Hansa was beaten to stop her from eating dirt.

The zoo was particularly thrilled to have her in part because in 2003 federal authorities started cracking down on imports of wild-born elephants to the United States. Without imports, the total number of elephants in the country started to decline. Zoo elephants were dying young—and dying faster than they were reproducing. Meanwhile, the public was beginning to ask questions about whether breeding elephants in captivity was a good idea. Berens reported that in 2005 the AZA held a meeting to address the elephant problem. The assembled directors of zoos with elephants decided to hire a crisis-management firm to help them fight back against anticaptivity activists that they dubbed "extremists." And the group recommitted to breeding elephants.

That same year, Hansa turned five. She was given cupcakes made of cornmeal, carrots, grapes, raisins, and bamboo leaves, with pumpkin frosting. By now she was 3,900 pounds, but still nursing. Elephants in the wild develop slowly and live as long as humans do. The zoo's head zookeeper told the *Seattle Times*, "Our hope is that she will stay here, and reproduce here, and that her babies will be here as well." The Woodland Park Zoo continued to artificially inseminate Chai, hoping for more babies. Ultimately, they would try more than 100 times, without success.

Then in 2007, little Hansa died—from herpes. The zoo announced the death at a press conference. Several of the staff wept. "Hansa was a symbol of hope and a true ambassador for her species," Woodland Park Zoo President Deborah Jensen said. "She deeply touched our lives and she inspired many of us to help elephants and other animals." Hansa's body was left in the enclosure for a time, which was closed, so the other elephants could mourn her.

The Woodland Park Zoo donated $8,500 to an elephant sanctuary in Borneo in honor of Hansa—a donation touted in an editorial co-written by a member of the zoo's board of directors as proof that "[e]very time you visit an elephant in an accredited zoo, a portion of your admission fee supports elephant conservation." The zoo's conservation director flew to Borneo to present the check in person. Given that the zoo brought in 6.5 million dollars in tickets and parking fees that year, it was a pretty small portion.

After the *Seattle Times* investigation was published, the Woodland Park Zoo responded with an op ed from two members of their board of directors, arguing that breeding elephants was part of conserving them in the wild. "We have learned lessons about elephant reproduction, communication and behavior that never could have been gleaned from wild populations." They defended their care of Watoto, Chai, and Bamboo. And, of course, they argued that the elephants were ambassadors for their kind, writing that "[s]eeing, hearing and smelling elephants can spark a very personal, emotional connection that inspires people to help elephants in the wild."

An independent report on the welfare of the zoo's three elephants in 2013 found them to be physically healthy but mentally under-stimulated, as evidenced by their repetitive weaving and rocking behaviors. The report also recommended replacing the concrete floors of their barn with sand, which would reduce stress on their feet and legs. Then, in 2014, forty-five-year-old Watoto collapsed. A medical report said her legs were extremely arthritic. "My clinical assessment is that she was unable to stand back up, due to the joint disease," said Darin Collins, a zoo veterinarian. "When lying down, large-bodied animals cannot breathe normally owing to massive weight impacting their lung cavity, decreasing blood flow to vital organs and nerves, and resulting in limb paralysis." She simply couldn't get up, and she was dying, so they killed her to end her suffering.

On cue, Martin Ramirez, the zoo's curator of mammals, reminded the public why captive elephants are okay: "We are very grateful to our community for the support they have shown us as we grieve the loss of Watoto. We will miss her regal presence and hope that people don't forget her and the role she played as a champion for her cousins in the wild."

Some Seattleites began to demand that the zoo's elephants be sent to a refuge, like the Elephant Sanctuary in Tennessee, or the Performing Animal Welfare Society (PAWS) in San Andreas, California—places organized for their happiness, where they would not be bred or displayed. Retired zoo elephants at PAWS take mud baths and swim in a lake. The Elephant Sanctuary doesn't even offer tours. But they do have a few webcams up. Clicking through the cams, looking for the elephants, I find a scene that gives me goosebumps. The camera is mounted high, high up, overlooking a massive lawn edged with trees. Shadows of clouds slide across the grass. Far, far back in the distance, I can just make out two tiny gray shapes: elephants with plenty of space to run at full speed, if they want to.

In Seattle, activists demonstrated and posted videos of Chai pacing and swaying, moving her gray bulk back and forth, back and forth, with dull eyes. But the Woodland Park Zoo faced an obstacle in the AZA, its accrediting institution. The AZA opposed elephant sanctuaries vehemently, since they took elephants out of the breeding pool. When zoos around the country started sending their older elephants to sanctuaries to retire, AZA threatened to—and sometimes did—yank their accreditation. (Today, the AZA has changed its tune on elephant sanctuaries. The organization even granted accreditation to The Elephant Sanctuary in Tennessee in 2017.)

Caught between souring public sentiment and the AZA, the zoo opted to transfer their elephants to the Oklahoma City Zoo in 2015. After the move, Chai began losing weight again, like she did during her time at the Dickerson Park Zoo, ultimately losing over 1,000 pounds. The Oklahoma Zoo had trained her to perform for the crowds, and during one of these performances Bamboo attacked her, knocking her into a fence.

On January 30, 2016, Chai was found dead in her cage. The cause of death was determined to be severe fat loss and a systemic blood infection. The *Seattle Times* found documents that showed that the elephant had struggled to get up twice in the weeks before her death, scraping her body against a wall, creating wounds that could have been the entry point for the bacteria. On the other hand, it could have been her teeth, which may have been chronically infected. "Unfortunately, they can't tell us if they don't feel well. We have to go on what we can detect," Oklahoma City Zoo veterinarian Dr. Jennifer D'Agostino told the paper.

Before Chai died, she had also likely transferred her herpes virus to a four-year-old elephant in Oklahoma, Malee, who had died of the infection.

Today, there are about 300 elephants at AZA-accredited zoos in the United States. But the scramble for new elephants continues. In 2016, three zoos—the Sedgwick County Zoo in Wichita, Kansas; the Dallas Zoo; and the Henry Doorly Zoo and Aquarium in Omaha, Nebraska—imported 17 elephants from a private game reserve in the Kingdom of Eswatini (formerly known as Swaziland). The reserve claimed it had no room for the elephants and would have to kill them if no home could be found for them. The zoos "donated" $450,000 for the animals. An investigation by *New York Times* reporter Charles Siebert showed that there was in fact, plenty of room for the elephants. So far, one of the imported elephants, Warren, age 8 or 9, has already died, when under anesthesia for surgery on a cracked tusk.

In 2019, the Convention on International Trade in Endangered Species of Wild Fauna and Flora tightened the rules on trading live African elephants. Exports from their native range are now only allowed in "exceptional circumstances" where there will be "conservation benefits for African elephants." It remains to be seen how these clauses will be interpreted by regulatory authorities.

Chai, Hansa, and Warren's experiences are hardly anomalies. Berens's investigation found that, in general, elephants in captivity in the United States died earlier and had much grimmer lives than their wild cousins. Their infant mortality rate—40 percent—was three times the rate in the wild. They suffered from "disease linked to conditions of their captivity, from chronic foot problems caused by standing on hard surfaces to musculoskeletal disorders from inactivity caused by being penned or chained for days and weeks at a time." Half of captive elephants were dying by age 23, even though elephants commonly live past 70 in the wild.

Zoos are not good homes for elephants. Aquariums are not good homes for cetaceans. But what about the rest of the menagerie? I think it is likely an increasing number of people will turn against keeping animals that usually have large territories, such as tigers and bears, and animals that are cognitively complex, such as apes. And I doubt the tide of public opinion will stop there. The new findings pouring in about animal intelligence suggest that lots and lots of species are sentient, intelligent, and can and do plan ahead. Seeing animals as having agency in this way makes their captivity in zoos tougher to take. If animals routinely make plans and take action in the world, being locked up stops them from acting freely. And surely one cannot flourish without freedom.

I asked Dan Ashe, the president and CEO of the AZA, if his organization had a plan to deal with declining public acceptance of animal captivity. He responded that he doesn't think it is declining. The AZA does quarterly surveys, he said, and "the people that have that strong anti-zoo, anti-captivity sentiment" represent about 8 to 10 percent of the population and that proportion hasn't changed over time. Elephants, he admitted, are among the "animals that carry a higher burden in terms of a social license," but new AZA standards were ensuring ever better care for zoo elephants, he said. For example, bullhooks are still used to train zoo elephants, although the AZA announced in 2019 that accredited zoos should "phase them out" by 2023. "The more we learn about caring for elephants, the better we'll be."

———

Many animals clearly show us that they do not enjoy captivity. When confined they pace, pull their hair, and engage in other repetitive tics. Researchers divide the odd behaviors in captive animals into two categories: "impulsive/compulsive behaviors," including "coprophagy [eating feces], regurgitation, self-biting and mutilation, exaggerated aggressiveness and infanticide," and "stereotypies," which are endlessly repeated movements. Elephants bob their heads over and over. Chimps pull out their own hair. Giraffes endlessly flick their tongues. Bears and cats pace. As many as 80 percent of zoo carnivores, 64 percent of zoo chimps and 85 percent of zoo elephants display compulsive behaviors or stereotypies.

Many zoos use Prozac and other psychoactive drugs on at least some of their animals. As an example, the Toledo Zoo has dosed zebras and wildebeest with the antipsychotic haloperidol to keep them calm, and has put an orangutan on Prozac. When their gorilla, Johari, began pulling out all her hair and fighting with the male she was placed with, Kwisha, she went on Prozac too. The drug kept her tractable. Instead of fighting off the big male, as she had been, she allowed him to mate with her, and—to the delight of zookeepers—she had a baby, Dara. Meanwhile, the male, Kwisha, would get so upset when any other gorilla was "immobilized for treatment or surgery," according to the *Toledo Blade,* that his keepers started giving him Valium before they sedated any of his cage mates.

Other animals actively try to escape. Jason Hribal's *Fear of the Animal Planet* chronicles dozens of circus and zoo animal escape attempts. Elephants figure prominently in his book, in part because they are so big that when they escape it generally makes the news. A surprising number of circus elephants have escaped or turned on abusive trainers—showing not only that they can tell humans apart, but that they reserve their animus for the humans that have done them wrong. Elephants, as they say, never forget. An Asian elephant named Mary was hit in the head by her trainer after a circus performance in 1916. She picked him up and threw him through a wall. Then she stepped on his head, killing him. Mary was hanged to death as punishment in front of a crowd of 3,000. Black Diamond was walking in a parade in Corsicana, Texas, in 1929 when he saw, in the audience, an old trainer who had been cruel to him. He snapped his chains and ran after the man, hurling him to the ground, breaking his arm, and killing the woman he was with. Janet burst out of the arena during a performance in 1992 with a mother and her kids on her back, then located and trampled people who had hurt her. Debbie and Frieda escaped together, twice, in 1995. Kamba escaped from the Texas State Fair in 1999 and had to be chained to a truck to be dragged back into captivity. Tonya escaped four times. In 1994, Tyke rebelled against her trainer, killing him, then escaped from a circus in Honolulu and was shot dead in the streets. An investigation determined that Tyke's "owners" were abusive and neglectful. Tyke's fellow elephants were seized and sent to sanctuaries.

Zoo elephants have also sought liberty—and revenge. Jumbo, a giant male African elephant who arrived at the London Zoo in 1865 used to hurl himself against the doors of his cage when in musth, the mating period, when testosterone floods a male elephant's body. He could only be kept calm with copious amounts of beer. In the 1910s, Babe escaped from the Toledo Zoo frequently, and would wander the neighborhood, eating flowers. He also gored a keeper to death, after which the zoo sawed off his tusks. In 1983, an elephant named Misty escaped from a zoo in Irvine, California, and killed a keeper there. In 1993, Tillie escaped the Lowry Park Zoo in Tampa, Florida, then killed a keeper a month later. In 1997, Cally and Tonya escaped from a zoo in Maine. The pair split up. When Cally was found hours later, she was taking a mud bath in the woods, a pleasure denied her in her zoo enclosure. In 2002, Moja crushed her handler at the Pittsburgh Zoo, and she and her three-year-old daughter Victoria took off. They were recaptured quickly. Elephants have a hard time staying escaped, given their size. Hribal found many more examples of zoo elephants hurting or killing their keepers and evidence that zoos routinely downplayed or even lied about these incidents.

And elephants aren't the only species that try to opt out of a zoo life. Tatiana the tiger, kept in the San Francisco Zoo, snapped one day in 2007, after a group of teenage boys had been taunting her. She somehow got over the 12-foot wall surrounding her 1,000-square-foot enclosure and attacked one of the teenagers, killing him. The others ran, and she pursued them, ignoring all other humans in her path. When she caught up with the teenagers at the café, she mauled two more before she was shot to death by police. Investigators found sticks and pine cones inside the exhibit, likely thrown by the three boys. Tatiana's case isn't unique. Other tigers and leopards are known to have escaped—and targeted their keepers or taunters.

Primates are excellent at escaping. Alphie, a Japanese macaque, escaped from the Pittsburgh Zoo in 1987 when the wind knocked a branch into his enclosure. Alphie crossed this bridge to liberty and was only recaptured six months later—in Ohio. Some escaped zoo monkeys were never recaptured.

Little Joe, a gorilla, escaped from the Franklin Park Zoo in Boston twice. At the Los Angeles Zoo, a gorilla named Evelyn escaped *seven times.* Apes are known for picking locks and keeping a beady eye on their captors, waiting for the day someone forgets to lock the door. One orangutan at the Omaha Zoo kept wire for lock picking hidden in his mouth. A gorilla named Togo at the Toledo Zoo used his incredible strength to bend the bars of his cage. When the zoo replaced the bars with thick glass, he started methodically removing the putty holding it in. In the 1980s, a group of orangutans escaped several times at the San Diego Zoo. In one escape, they worked together. One held a mop handle steady while her sister climbed it to freedom. Another time, one of them, Kumang, learned how to use sticks to ground the current in the electrical wire around her enclosure. She could then climb the wire without being shocked. Orangutans in Fresno spent weeks unraveling a nylon netting so they could slip through—but they never worked at their task when the keepers were around.

It is impossible to read these stories without concluding that these animals *wanted out.* Hribal researched dozens of escapes or attacks and found that zoos have a pretty standard PR playbook when they occur. First, they claim that such events are "rare" and then they deny the agency or intentionality of the animal, suggesting it was acting "on instinct." Hribal rejects this. These animals, he says, have been well trained by their keepers, sometimes with snacks and sometimes with bullhooks, electric prods, or other negative reinforcements. They know what they are supposed to do and they know that breaking the rules means no treats—or worse. Yet they bolt anyway. They are escaping intentionally, Hribal says. "They have a conception of freedom and a desire for it."

In at least one case, an animal has told us directly that he wanted out. Nim Chimpsky was a chimpanzee who was raised in a human household like a human child as an experiment and had learned sign language. But by age 10, he had ended up alone in a cage at a sanctuary in Texas. When he received a visit from a scientist he used to work with, he immediately signed "Bob," "out," and "key."

————————

Not all zoo animals actively try to escape. Many seem outwardly content. But even if their lives are pleasant, they may be short.

In 2014, the Copenhagen Zoo shot and killed a giraffe named Marius. It is likely that the public wouldn't have even noticed this—zoos euthanize "surplus" animals all the time—except that the Danes set off an international scandal by autopsying Marius in public as an educational exercise and then feeding what was left to the lions. Bengt Holst, the Copenhagen Zoo's scientific director, told the *New Yorker*, "It's fine to get attached to the animals—as our keepers are. But you also have to be realistic. This is not a fairy tale, where everything gets born but never dies."

Marius was surplus to requirements because most zoos keep only one male in their captive giraffe herds. Otherwise, with females around, the males would fight. Marius's father was already becoming aggressive toward him. At the time of his birth, there were simply too many male giraffes in the European zoo system. Since Marius's genes weren't needed for further breeding, there was no room for him.

(A surplus of males is a problem in many species, because of male aggression. In elephants, males have to be separated from each other, which can be extremely costly for zoos, given the size of the facilities required. But since even males are important to the breeding effort, an attempt was made to create a place to stash them. In 2013, The National Elephant Center was opened in Florida for that purpose, but it closed just three years later after three elephants in the prime of life died there.)

Instead of giving the breeding-age giraffes birth control, European zoos let them breed, since contraception can threaten some animals' fertility and since breeding makes animals happy. They then cull superfluous offspring. The European Association of Zoos and Aquaria (EAZA) kills between three and five thousand animals a year. During the coronavirus pandemic, the Neumünster Zoo in Northern Germany coolly announced plans to cope with lost revenue by feeding some animals to other animals, compressing the food chain at their zoo like an accordion, until in the worst-case scenario, only Vitus, their polar bear, would be left standing.

Getting information on how many surplus animals there are in the AZA network and what happens to them is difficult. AZA zoos apparently do kill surplus animals, but they don't cull large mammals as often. "We have a very strict detailed animal disposition policy within AZA," said president and CEO Dan Ashe. "It does allow for the euthanization of animals. But it's very rarely employed by our member institutions." Mostly, he said, they use contraceptives to prevent surplus babies being born, and when there are extra animals, they place them with other AZA-accredited or non-accredited facilities. "We have a cooperating partner framework where we look at them, we look at their animal care. We're not sending animals to places that we don't feel like we can support," Ashe said. It might seem like surplus animals should be released to the wild, but in most species, captive-raised animals can't hack it in the complex and dangerous world outside the zoo without intensive training.

Zoos accredited by the EAZA or the AZA have studbooks and genetic pedigrees and carefully breed their animals as if they might be called upon at any moment to release them, like Noah throwing open the doors to the Ark, into a waiting wild habitat. This isn't likely for most species. If zoo populations were really primarily "backup" populations intended for eventual release, then why would they be spread all over the world? Why wouldn't all the lions and giraffes be in African zoos, all the polar bears in Arctic zoos? Nevertheless, the keepers of the studbooks try to keep genetic diversity high, and frown upon mixing up subspecies, or hybridization of any kind. They are, after all, respectable conservation-oriented institutions. But that day of release never quite seems to come.

There are a few exceptions. The Arabian oryx went extinct in the wild and then was reintroduced from zoo populations. The California condor breeding program, which we will look at in-depth in the next chapter, was hosted at zoos. The AZA says that its members host "more than 40 reintroduction programs for species listed as threatened or endangered under the Endangered Species Act." Nevertheless, the vast majority of zoo animals (800,000 animals of 6,000 species in AZA zoos alone) will die in captivity, whether culled as surplus or dying of old age after a lifetime of display. Tellingly, of the peer-reviewed research articles coming out of AZA-accredited zoos and aquaria, only 7 percent focus on species conservation. The bulk of the research is on the biology of the animals and their veterinary care.

Ashe suggested that learning how to keep and breed wild animals in captivity can be thought of as a kind of conservation research. While conserving populations in the wild is everyone's first choice, a day may come when the only option is captive breeding. And so, he says, we need to know how to do that. "If you don't have people that know how to care for them, know how to breed them successfully, know how to keep them in environments where their social and psychological needs can be met—then you won't be able to do that."

The Department of Defense, he pointed out, doesn't build battle readiness after war is declared. They stay ready. "We were in a war to conserve biological diversity," he said. "We need to decide what is the capacity that we're going to need in order to win that war, and build it. And the fortunate thing for things like elephants and great apes and polar bears is we have an industry that wants to do it and that serves the public in important ways." That justifies, for Ashe, keeping rare or endangered wild animals. But keeping common wild animals, as long as they are well cared for, doesn't bother him either.

"I don't see any problem with holding animals for display," Ashe told me. "People assume that because an animal can move great distances that they would choose to do that. Their walking, their movement is purposeful; they're moving for water, for food, for safety, to find a mate." If they have everything they need nearby, he argued, they would be happy with smaller territories. And it is true that the territory size of an animal like a wolf depends greatly on the density of resources and other wolves. But then there's the pacing, the rocking. I pointed out that we can't ask animals if they are happy with their enclosure size. "That's true," he said. "There is always that element of choice that gets removed from them in a captive environment. That's undeniable." Ashe just doesn't object to captivity per se, as long as the conditions for the animals are great and the animals are healthy and content. His justification was surprisingly philosophical. In the end, he says, none of us are free. "We are all captive in some regards to social and ethical and religious and other constraints on our life and our activities."

———

What Ashe wants visitors to experience when they look at the animals is a "sense of empathy for the individual animal, as well as the wild populations of that animal." In 2011, the Woodland Park Zoo introduced wolves into its new "Northern Trail" exhibit, four one-year-old females from the New York State Zoo. Zoos don't need to breed wolves for conservation. Although many people think of them as threatened because they were long absent from much of the American West, globally they are doing fine. Alaska, Canada, and Russia host hundreds of thousands of wolves. The International Union for Conservation of Nature (IUCN)—a nongovernmental organization that maintains an influential "red list" of endangered species—categorizes them as "least concern." So the argument, then, is the old "ambassador" gambit. "The wolves are conservation ambassadors representing the complex and volatile story of the return of the wolf to Washington State and the challenges their endangered cousins in the wild face," the zoo wrote in a blog post introducing their new animals. How four bright-white Arctic wolves (not a color found in Washington State) being cooped up in a tiny enclosure in Seattle "represented" any of those things was left a little vague. "Given the rising political pressures and increasing conflict between wolves and people in the Northwest and Northern Rockies, it's important for people of all ages to connect with wolves at the zoo and learn about the challenges these predators face in the wild, the unwarranted fears and their contribution to our ecosystems." The argument being made here is that by seeing real live wolves—perhaps howling at them for ten seconds before moving on to the next exhibit—people will be more likely to support the reintroduction of wolves to the American West.

I do not doubt that some people had their passion for a particular species, or wildlife in general, sparked by zoo experiences. I've heard and read some of their stories.

Once, I overheard two schoolchildren at the National Zoo in Washington, DC, (formally known as the Smithsonian's National Zoo & Conservation Biology Institute) confess to one another that they had assumed that elephants were mythical animals like unicorns before seeing them in the flesh. I remember well the awe and joy on their faces, 15 years later. I'd like to think these kids, now in their early 20s, are working for a conservation NGO somewhere. But there's no unambiguous evidence that zoos are making visitors care more about conservation or take any action to support it. After all, 200 million people visit a zoo every year and biodiversity is still in decline.

Researchers quizzed visitors to the Cleveland, Bronx, Prospect Park, and Central Park zoos about their level of environmental concern and what they thought about the animals. Those who reported "a sense of connection to the animals at the zoo" scored higher on environmental concern, as did zoo members as compared to irregular visitors. On the other hand, the researchers reported, "there were no significant differences in survey responses before entering an exhibit compared with those obtained as visitors were exiting." The team concluded that "zoo experiences may contribute to the development of an environmental identity over time among zoo members" but that zoos are "primarily valued as a recreational opportunity for visitors: a place to relax and also to promote social interaction and family togetherness."

A 2008 study of 206 zoo visitors by the same team showed that while 42 percent said that the "purpose" of the zoo was "to teach visitors about animals and conservation," 66 percent said that *their* primary reason for going was "to have an outing with friends or family," and just 12 percent said their intention was "to learn about animals."

The same research team also spied on hundreds of visitors' conversations at the Bronx Zoo, the Brookfield Zoo outside Chicago, and the Cleveland Metroparks Zoo. They found that only 27 percent of people bothered to read the signs at exhibits. More than 6,000 verbal comments made by the visitors were recorded, most of them variations of "Hey, look at that!" People looking at primates tended to talk about how similar they are to humans. The researchers wrote that "In all the statements collected, no one volunteered information that would lead us to believe that they had an intention to advocate for protection of the animal or an intention to change their own behavior."

I interviewed conservation psychologist John Fraser, who was a co-author on both of these studies, and told him my reading of his results was that zoos were not making much of a difference. He disagreed. A single visit to a zoo, he said, isn't going to make a person "suddenly have an epiphany, sell their SUV, and start living exclusively on nuts that fall from trees." But a series of visits and, perhaps even more importantly, the way memories of zoo visits linger, can shape a person in subtle ways. "Conservation is a value system," he said. "You don't teach values and morals in one shot." It is extremely difficult to tease out and prove the role of a series of childhood and adult zoo visits in creating a person who cares about wild animals and the environment—let alone calculating an average effect. I wanted a study that said something like "10 zoo visits makes a person 10 percent more caring about the environment." Fraser said I was never going to get it.

His years of experience studying zoos have convinced him that they do increase people's moral concern for the non-human world. This effect is not always intentional, however. He suggested I check out his dissertation, in which he interviewed working conservation biologists about early zoo experiences. Many of his interviewees, looking back at zoo visits in the mid-20th century, remembered a moment of clarity in which they decided that a zoo was *not* properly caring for an animal or that animals maybe didn't belong in zoos at all. "Zoo animals were particularly relevant to the development of environmental identity when they were individualized as personalities or recognized as expressing individual agency against negative conditions in a zoo," he wrote. "[I]t appears that some conservation biologists were provoked by negative experiences that were not representative of good animal care, while the positive experiences did not emerge as transformative learning experiences."

————

People don't go to zoos to learn about the biodiversity crisis or how they can help. They go to get out of the house, to get the kids some fresh air, to see interesting animals. They go for the same reason people went to zoos in 1828: to be entertained.

In some ways, it is kind of reassuring that people will still leave the house and pay money to see real animals in person in an era of Animal Crossing, the CGI remake of *The Lion King*, and glossy nature documentaries narrated by David Attenborough. It means that people still like real animals and it means that seeing real animals in person still scratches some primeval itch in a way that those more mediated, more virtual experiences do not. But that special feeling is even better, brighter, and more memorable when people see real animals *outside* of zoos, in their normal habitats.

I've seen dozens of large and fierce animals at zoos in my life—tigers, hippos, emus, elephants—and I do have a few individual memories of these experiences. In 2006, I watched hippos at the National Zoo in DC eat entire bunches of bananas like they were Tic Tacs. But mostly my memories are a blur of fur, ice cream, and an increasing sense of uneasiness about the lived experience of the animals I was gazing at. In contrast, I have searingly vivid memories of seeing a badger run across a road in Eastern Oregon, of a coyote with a duck in its mouth at a wildlife refuge, of spotting sandhill cranes in a corn field, of a muskrat swimming under the ice in a canal near my house, catching a quick breath in an air bubble trapped under the ice's surface, of watching pelicans soar just a few feet above a wetland through which I was canoeing. These are not high-dollar, pack-'em-in-the-door animals that zoos put on their T-shirts. But our encounters were unexpected, magical, lucky. Each time, I felt excited to glimpse animals that are hard to see. But I was also interested in what they were doing—what they were *choosing* to do. The badger decided to cross that road; the coyote claimed that duck for its dinner; the muskrat figured out how to stay safe and cozily underwater during a freeze. I was looking at animals leading their own lives, not animals whose lives were devoted to my viewing pleasure. These were real *encounters*. SeaWorld's "Orca Encounter" is not—the orcas have no choice but to be there.

What if zoos stopped breeding all their animals, with the possible exception of any endangered species with a real chance of being re-released into the wild? What if they sent all the animals that need really large areas or lots of freedom and socialization to refuges? With apes, elephants, big cats, and other large and smart species gone, they could expand enclosures for the rest of the animals, concentrating on keeping them lavishly happy until their natural deaths. Eventually, the only animals on display would be a few ancient holdovers from the old menageries, some animals in active conservation breeding programs, and perhaps a few rescues.

Such "zoos" might even be merged with sanctuaries, places that take wild animals that—because of injury or a lifetime of captivity—cannot live in the wild. Existing refuges, like Wolf Haven, often do allow visitors, but not all animals are on the tour, just those who seem to like it. Their facilities are really arranged for the animals, not for the people. These refuge-zoos could become places where animals live not in order to be on display, but in order to live. Display would be incidental.

Such a transformation might free up some space. What could these institutions do with it, besides enlarging enclosures? As an avid fan of botanical gardens, I humbly suggest that as the captive animals retire and die off without being replaced, these biodiversity-worshiping institutions devote more and more space to the wonderful world of plants. Properly curated and interpreted, a well-run garden can be a site for a rewarding "fun day out," a source of education for the 27 percent of people who read signs, and a point of civic pride. I've spent memorable days in botanical gardens like the Fairchild Tropical Botanic Garden in Coral Gables, Florida, the Missouri Botanical Garden, the New York Botanical Garden, and the Minnesota Landscape Arboretum, completely swept away by the beauty of the design as well as the unending wonder of evolution—and there's no uneasiness or guilt. When there's a surplus, you can just have a plant sale.

Convincing the boards and executives of zoos to wind them down, transition them to refuges, or transform them into botanical gardens will be tough. There's institutional inertia to overcome. And there's psychology. Most people who run and work at zoos are there because they genuinely adore animals, and they see themselves as the loving caretakers of the creatures in their charge. In many modern zoos, animals are well cared for, healthy, and probably, for many species, content. Zookeepers are not mustache-twirling villains. They are kind people, bonded to their charges, and immersed in the culture of the zoo, in which they are the good guys. This was one of the most remarkable lessons of *Blackfish*. The "trainers" who worked with the captive orcas stayed at SeaWorld because they felt a duty of care to the animals. They thought to themselves, "If I leave, who is going to take care of Tilikum?"

I don't think breeding animals for a life in zoos can be defended in any ethical system. It violates the rights of many animals to express their capabilities and flourish. It lacks compassion. It is the wrong sort of "care." Even the utilitarian argument falls short, since so little "good" is produced by the animals' exhibition. My preference for a "fun day out" does not justify generations of animal captivity.

I mentioned that my grandmother, Jean Beck, used to take my brothers and me to the Woodland Park Zoo frequently, back in the 1980s. She even had a membership. I remember walking to the zoo with her from my house, holding onto the handle of my brother's stroller while heading up busy 50th Street under a canopy of big old maple trees. I remember the sweetness of the red sugar water at the bottom of the sno-cones she would buy for us. And I remember seeing elephants, but I don't remember their names or really understanding them as individuals. While writing this chapter, I talked about those days with my mother. She surprised me by telling me that my grandma, who died at age 92 in 2015, took us kids to the zoo because we loved it, but that she, in fact, hated it, and hadn't been for decades until we asked to go as children.

My grandmother grew up just a few blocks from the zoo in the 1920s and '30s. Back then, there were no gates around it; it was free. So she used to cut through on her way home from school. One cold, rainy day, on her way home from Lincoln High School, she went past some caged monkeys. The monkeys looked miserable. According to my mother, she said to herself, "That's it." And she never walked home through the zoo again.

7

The Dignity of the Condor

Wild pets and zoo animals are prisoners of our desire to have good relationships with non-human animals. Yet our control over them makes it impossible for the relationship to be mutually respectful—to be good. Their captivity is unjustifiable because it is *for us*, to satisfy our whims. But there are other reasons we capture or meddle with the lives of our non-human kin. Often, humans, in the shape of conservation biologists or wildlife managers, swoop into the lives of individual animals to protect the diversity of life on Earth. I believe protecting biodiversity is a worthy goal. It is extremely important to me. But I'm beginning to wonder if there are limits to what we should do to achieve it.

Let's start with the captive breeding of rare animals for reintroduction into the wild. In some ways, captive breeding facilities are like zoos. Indeed, they are often located at zoos. But these programs aren't breeding animals for humans to look at, with the idea that their descendants could also function as a worst-case-scenario backup for wild populations. These programs are usually eager to get their animals back out onto the landscape as quickly as possible—ideally before they begin to adapt to captivity. And yet, like zoos, the programs are not really *for* the individual animals involved. They are *for* their species. The difference is subtle but it is the key to this ethical puzzle. Does saving the kind justify restricting the autonomy of the individual?

———

Their wings are dusty black, a white stripe underneath, the primary feathers individually distinguishable against a hard blue sky. Their naked coral heads sport a serene, watchful expression above an elegant, fluffy feathered collar. They soar as if gliding on ice. California condors—fantastically large wild birds with a nine-and-a-half-foot wingspan—once flew over most of North America. During the Pleistocene extinctions, the disappearance of the world's megafauna (and their mega-carcasses) killed off seven other kinds of vultures. Lucky for the California condors living through those years, there were beached whales and sea lions to bury their beaks in.

By the time Europeans arrived in North America, the species' range had shrunk, extending from the Pacific coast at least as far east as the Cascade and Sierra Nevada mountain ranges, and bounded by British Columbia to the north and Baja California, Mexico, to the south. Native people up and down the coast held the condor in high regard and there are records and artifacts showing its place in the spiritual lives of the Ohlone, Miwok, Maidu, Mono, Chumash, Yokuts, and Luiseño peoples. The Yurok people, who live in the coastal redwood forest and along the Klamath River in what is today called Northern California, call the California condor *prey-go-neesh*. They use its feathers in the World Renewal ceremony, during which they rebalance the world through fasting and prayer. But by the middle of the 20th century, prey-go-neesh no longer soared above Yurok territory. Its range had shrunk to just the San Joaquin Valley of California and the surrounding region. There were lots of things killing the birds, but humans were among them—both directly and indirectly. Hunters shot the birds outright, but they were also being poisoned by the carcasses of predators, such as bears and wolves, and other "vermin" that humans had killed with strychnine and other poisons. Another problem hidden for decades was the lead shot that European settlers used in their guns. The metal was deadly poison to the birds, who perished from eating carcasses left riddled with the stuff.

The role of lead in the bird's decline wasn't immediately clear, and for many years protected areas were established to maintain its habitat—arid, rugged terrain, with plenty of cliffs upon which to nest—with little effect. Condors have only about one chick every two years, and they were dying faster than they were breeding. By the late 1970s, groups such as the American Ornithologists' Union and the National Audubon Society began calling for captive breeding of the birds.

But in 1981 the environmental group Friends of the Earth published a book called *The Condor Question: Captive or Forever Free?*, which argued against any captive breeding program. Prominent conservationists—including Anne and Paul Ehrlich, whose 1968 bestseller *The Population Bomb* had forecast an imminent global food shortage apocalypse that never came to pass—opposed the project. The Ehrlichs called the "space-age technology intended to 'save'" the species "outrageous" and predicted it would "*hasten* the extinction of the California condor." *The Condor Question*'s contributors had a number of practical objections to the plan, but ultimately their concern was philosophical. A cage-raised condor wasn't just worth less than a wild condor, in their view. It wasn't a condor at all. "A wild condor is much more than feathers, flesh, and genes," wrote condor expert Carl Koford; "A cage-raised bird can never be more than a partial replicate of a wild condor . . . A condor in a cage is uninspiring, pitiful, and ugly to one who has seen them soaring over the mountains." David Brower, founder of Friends of the Earth, wrote that condors had rights, including "the right to their own dignity," and "freedom of the sky."

Most of the book's writers were optimistic that the bird could be saved without captive breeding. But experts writing elsewhere were willing to admit that the bird might be lost. As renowned birder Richard Stallcup put it, "Must we burden and demean the doomed skymasters with electronic trinkets, then imprison them in boxes and demand that they reproduce? Or can we just say, 'Yes, el condor, we blew it long ago, we're sorry. Fly, stay as long as you can, and then die with the dignity that has always been yours.'"

With the estimated population in the wild down to fewer than 30, a consortium of government agencies and private groups decided to act—never mind the dignity of the birds. They captured their first nestling in 1982. In 1986 the National Audubon Society, which had apparently changed its mind about the captive breeding scheme, sued to stop the recovery team from taking birds from the wild. That December, about 60 people showed up at the Los Angeles Zoo to protest the plan to capture the last remaining wild birds. Some of the protestors dressed in beaks and giant wings.

The lawsuit and protests were unsuccessful at stopping the roundup of the birds and by 1987 the recovery team had every single remaining California condor but one captured. The last free bird, "Adult Condor 9" or "AC-9" had seen his friend nabbed by a remote-controlled cannon net in February and was wary. He was finally captured on Easter Sunday, 1987. A grainy film of the moment shows the giant bird alighting on the ground, then getting covered by a net that bursts out of the ground, along with a whole passel of what look to be ravens. A man in a ball cap pops out of a foxhole and sprints over to him and puts his hand on the struggling bird, as if to calm him. Then they load the bird into a dog kennel and pop it in the back of their truck.

The original plan was to take AC-9 to the Los Angeles Zoo, but protestors had chained themselves to the entrance, so he went to the San Diego Zoo instead, and transferred to Los Angeles later. As a *Sports Illustrated* article from the time put it, AC-9 had entered "a scientific twilight zone of computerized gene-tracking, puppet-mothering and behavior modification." That was it, the whole species was in a couple of zoos—just 13 males and 14 females.

Brower lamented to an interviewer that "the condor was destroyed in order to save it." He accused zoos of wanting the birds for display, a charge the San Diego and Los Angeles zoos denied. In defense of those who opposed the capture of the condors, there were many unknowns. Would capture prove so traumatic that many of the precious birds would perish? Would the birds reproduce in captivity? If they did, could their offspring learn the skills they needed to survive in the wild while still in aviaries?

The captive birds were deloused with a pesticide—thus causing the extinction of the California condor louse, *Colpocephalum californianus*. They were paired off and assigned aviaries. The birds could no longer choose where to live or with whom to mate. Their mates were chosen based on genetic dissimilarity, to avoid inbreeding. Their first clutch of eggs were taken from them so they would breed twice in a season, and their first round of offspring were raised by conservation biologists wearing condor puppets on their hands.

Do condors mind captivity? One unexpected source of insight is a 1910 report, written in slightly archaic turn-of-the-century English, by Oregon conservationist William Lovell Finley, which hints that being penned up could be a negative experience for the giant birds. "Altho the California Condor in the wild state likely lives quite a long time, the bird in captivity as a rule does not survive many years, nor does it have the brilliancy of color found in the wild specimens," Finley wrote. The author goes on to recount his experience raising a condor himself for no apparent reason other than curiosity. A nestling was kidnapped and taken to Portland, Oregon. "We placed General, as we called him, in an enclosure of about twelve by fifteen feet that we made under the trees. We gave him the stump of an old apple tree to perch on, but the primordial freedom of his race lingered within him, for he did not like the idea of being closed in." Thankfully, Finley allowed him out every day to "take his bath in the creek and warm himself in the sun." Apparently, General could tell when his outing was due. "If he were not release at the usual time, he became restless and soon attracted our attention by climbing up and poking his nose thru at the gate. The minute I opened it he stalkt out, but always stopt cautiously a moment or two outside the gate to look about. He did nothing without deliberation. With several hops he went half way across the yard, flapping his big wings. Then he went thru a regular dance, as if celebrating his freedom."

Finley found General to be both intelligent and affectionate, and clearly loved him. "One might think a person could have little attachment for a vulture. There is nothing treacherous or savage in the condor nature . . . We had fed him by hand on small bits of raw meat, from the beginning, and he showed an intelligence that was as markt as in any pet we have ever had. He loved to be petted and fondled. He liked to nibble at my hand, run his nose up my sleeve, and bite the buttons on my coat, and he was gentler than any pet cat or dog."

Mike Clark, an animal keeper at the Los Angeles Zoo, developed a similar attachment to AC-9. He started working with him in 1989 and grew particularly fond of the bird, whom Clark described as "very stout, compact. I always said he looked like a football player." What I wanted to know was whether AC-9 suffered, psychologically, in captivity. Did he try to get out? Did he pine for wide-open skies? According to Clark, all condors apply their fine minds to the problem of getting out of their enclosures at first, but eventually they seem to accept their new homes. In the many years he knew him, Clark recalled, AC-9 seemed content in his "flight cage," a chain-link enclosure against a canyon wall about 100 feet long by 40 feet wide. (Large enough to fly in, unlike General's much smaller pen.) "AC-9 was a very happy bird," Clark said. He said it's pretty easy to determine whether the birds are happy or not, once you spend some time with them. "You can tell by the expressions on their face, the way they sun themselves, the way they go and lay down and relax or bathe themselves or do some mutual preening."

Happy birds are also more likely to play. Each enclosure includes a perch on a scale so the keepers can just peek at the dial to check the weight of any bird that lands there. But the scale also has a little spring to it, which made it a fine toy for AC-9. "He would bounce up and down and jump, and peck at the moving needle," Clark said. "Just playing."

All captive breeding programs have to wrestle with the fact that captivity very quickly selects for the ability to tolerate captivity. Individuals that chafe against confinement or the particular conditions they are kept in will die or fail to breed. Those who can cope and reproduce have babies that share their ability to live in captivity. To avoid this, conservationists are always itchy to get captive animals back outside. Condors did better in cages than anyone expected, but their human caretakers were still eager to return them to their ecosystem.

The first condors were released in 1992, but after their time inside they were attracted to human structures and several were killed flying into power lines. The remaining birds were re- captured. Biologists decided that before they could attempt any further releases, they had to train the birds to stay away from people and buildings. The birds were allowed to interact with fake utility poles that gave them mild electric shocks, strong enough to make them want to avoid them in the future. Eventually, releases resumed.

As of March 2020, there are about 200 condors free in California, 90 in Arizona and Utah, 39 in Baja, Mexico, and 181 still in captivity. You can see one condor couple, Siwon and Sola, and their son Tiyep via a webcam set up in their enclosure at the California Condor Breeding Facility at the San Diego Zoo Safari Park. For a mere $58, paying guests to the San Diego Zoo can also see condors in person. The zoo's website advertises "up-close views of condors not in our breeding program at the Park's Condor Ridge and in the Zoo's Elephant Odyssey." Condors are also on display at the "Condors of the Columbia" exhibit at the Oregon Zoo in Portland, and at the Santa Barbara Zoo. I guess David Brower was right about at least one thing.

The condors that have been released into the "wild" are still tended to pretty closely by humans. They are routinely vaccinated against West Nile Virus. When chicks are in the nest, they are visited monthly to make sure their parents aren't feeding them plastic trash. If they are, the nestlings are whisked away for a quick surgery to remove the plastic. Every condor is assigned a "studbook number," which it wears prominently on a wing tag. "When you see a condor flying in the wild, look for the number printed on its wing tags," the Fish and Wildlife Service website recommends.

Many wear radio collars and are fed supplemental food to keep them away from carcasses that might contain lead shot. Every year, dozens of condors are captured to be tested for lead poisoning. Those who are sick are treated. Alas, many still die from lead poisoning. It wasn't until July 2019 that lead shot was fully outlawed in California, after some 76 recorded condor deaths since 1992.

"If they weren't being poisoned regularly, we could take the tags and transmitters off and be done with it," Clark said. "It is the only reason there was a program and why there continues to be a program." Even after the ban, poisoned birds are showing up. The very day I called to speak to Clark in June 2020, a female condor had been admitted for treatment. "She was just released in the fall, and she was 20 pounds, and today she is 12 pounds and has lead poisoning and is 90 percent going to die," Clark said. "They are dying almost as fast as we can release them."

It doesn't take very many people flouting the lead ban to endanger the birds. "We had a situation in the Grand Canyon where there was a carcass that someone was just out there blasting away at it with a shotgun," Clark said. "They were just shooting it for fun. We lost ten birds on that one thing."

Since mortality in the wild is higher than birth rates in the wild, the population remains supported by periodic releases of captive-raised birds. The species is not self-sustaining in the wild. But it is no longer on the verge of extinction. The Yurok Tribe Wildlife Program and the Fish and Wildlife Service are working planning on reintroducing prey-go-neesh to Yurok territory as early as 2021. "He carries our prayers to the heavens being the animal that flies the highest in our region," Tiana Williams-Claussen, a Yurok biologist and tribe member told the *Medford Mail Tribune*. "It's pretty significant."

Such a release could mean condors flying over the Rogue River Valley, just across the mountains from where I live in Southern Oregon. We once had condors here in the Klamath basin too. I can't say that the idea of seeing them return isn't exciting.

AC-9 was released at the Hopper Mountain National Wildlife Refuge in 2002, a mature bird who had spent most of his life in captivity. Clark said the moment was "bittersweet." He and his colleagues were keen for the bird to transmit his knowledge of the wild to a younger generation, but they knew the outside world was risky. He returned to his original territory, found a new mate, and built a nest in the very same nest cavity he had used in the 1980s. His movements were constantly monitored, but he was free.

On June 30, 2016, his transmitter stopped sending data. His last known location was "a remote canyon in the Sespe Wilderness," according to the Fish and Wildlife Service. Clark's best guess is that AC-9 ate a carcass contaminated with lead shot, got sick, went home to his nest, died in the nest, fell out, and was scavenged by another animal. Hopper Mountain staff eulogized him, writing that he would "be remembered for his many contributions to California condor recovery; first as one of the 22 survivors and then as a prolific breeder in captivity, producing 15 chicks from 1991 to 2001."

Adult Condor 9 didn't just contribute to California condor recovery; he sacrificed for it. He gave up his first mate, his territory, and his liberty, along with his ability to choose his own future mates and daily activities. On the other hand, if conservationists hadn't nabbed him with that net back in 1987, he would probably have died of lead poisoning while still a young bird. "When you have treated as many lead birds as I have, you watch them wither away and die, I guarantee you that they would rather be in captivity," Clark said. "Because it is miserable out there." In this case, the breeding program probably saved the individual birds as well as the species.

In 1981, environmental thinker John Farnsworth read *The Condor Question* and was "entirely persuaded by it." More than 30 years later, Farnsworth, now a Santa Clara University professor, reread the book, then visited the California Condor Field Station in Baja California, Mexico. The birds weren't really wild, in his opinion. They dined on lead-free meat provided by humans and wore tags that transmitted their whereabouts. Nevertheless, Farnsworth found the birds "glorious" to behold, "statuesque, unexpectedly dignified, and magnificent in almost every regard." He described the audible *whomp* when they touched down, and the intelligent curiosity with which they explored the world.

In retrospect, Farnsworth says, the trapping and monitoring and meddling that the authors of *The Condor Question* objected to so much turned out to be essential to discover the central role of lead in the bird's decline. The captive breeding program almost certainly did save the species from extinction. The predicted problems of captive-reared birds adjusting to the wild did appear, but proved to be speed bumps, not roadblocks.

"The birds I observed in Sierra San Pedro Martir were not mangled facsimiles of condors, nor were they only half the birds they should have been. Yes, they still rely on conservation efforts, and they will need to be monitored, protected, and perhaps even coddled well into the future. But they are a vibrant community, learning as they go how to become wild creatures, teaching each other how condors are supposed to behave."

It seems that all those concerns about dignity and wildness were really human concerns, not condor concerns. "Like most birds, they appear disengaged about existential issues such as extinction," Farnsworth writes, "preoccupied by the more immediate issues of feeding and breeding."

Condors are not the only species we've taken into protective custody. Humans have embarked upon extraordinary projects to save dozens of species of animals, from jewel-colored tropical frogs to everybody's cuddly favorite: the panda bear. With less than a couple thousand in the wild, the hundreds of pandas in captivity are cajoled to mate, ogled by zoo visitors, and—in a sign of our deep ambivalence about intervening in their lives— tended to by humans in panda costumes, scented with *eau de* panda pee, so that the bears won't become habituated to people.

Before considering captive breeding, conservationists are likely to try other, less intensive approaches, starting with protecting habitat, controlling threats to the species, or perhaps translocating them to better or safer habitat. Captive breeding is an expensive, emergency measure that is typically only undertaken when a species looks like it will go extinct without help. In order to succeed, captive breeding programs must not only house and breed the animals they are trying to conserve, but also secure adequate habitat into which to reintroduce the species, and eliminate or engineer solutions to ongoing threats, such as poachers, toxins in the environment, or introduced predators. Only when a self-sustaining wild population exists can the project truly be called a success. The condor project is still struggling with the second half of its mission, but other projects have succeeded in seeding populations that went on to thrive in the wild. Peregrine falcons and Arabian oryx are back and reproducing on their own now, after getting a helping hand from humans.

Sometimes conservation scientists even reach over the cliff edge to yank species back from the abyss. In the 1960s, some 2,300 northern white rhinoceroses lived in the wild. Today, thanks mostly to poaching for rhino horn, there are only two northern white rhinoceroses left, and they are both female. Many will remember an arresting photo by Ami Vitale of Joseph Wachira, a staffer at the Ol Pejeta Conservancy in Kenya, posing with Sudan, the last male northern white rhinoceros. In the image, Wachira lays his hand tenderly on the great animal and presses his forehead against

Sudan's as the rhino dies. You would think this would mark the end of the line for this subspecies, but no.

Scientists at the San Diego Zoo Institute for Conservation Research, led by geneticist Oliver Ryder, have a freezer full of northern white rhino cells, including cells from Sudan, and they have figured out how to coax the cells to turn into sperm. I visited the project in the spring of 2019, and was shown the cells under a microscope, wriggling around, tails and all. The idea here is that these sperm could be used to inseminate the remaining females, or females of the related southern white rhino, of which the San Diego Zoo has a small breeding population. "We are in the era of bold intervention," Ryder told me.

I asked him how different the northern and southern white rhinos were from one another. Their "native" ranges are separated by hundreds of miles, but they aren't easy for lay people to tell apart. He said they had diverged only 15 to 30 thousand years ago. I was stunned. That's *nothing* in geological time. Was it really necessary to keep these lineages separate? Even if Ryder's team is able to create some new northern white rhinos, they will have a tiny gene pool. But for Ryder, it is the principle of the thing. "We can choose to prevent the extinction of the northern white rhinoceros or we can let that happen," he said. "Are we just going to turn our backs and say, 'No, it's too hard?' This is my way of giving back. If we can figure out how to save the northern white rhino, it will make me feel better about what we have done to the planet."

Captive breeding is in some respects astonishingly bold, almost brazen. In order to create the many releasable individuals that are the ultimate goal of the exercise, we must maintain several generations of captive individuals, keeping them healthy and, ideally, providing them with a life that has opportunities for joy and meaning—like the "generation ships" in science fiction that take colonists on centuries-long journeys to new planets. We must do all this despite the fact that we cannot communicate with them directly, cannot ask for consent, cannot really know their lived experience. It's an exercise in total domination, undertaken as part of a larger cultural project of stopping extinctions, which is arguably an attempt to reverse or reduce human domination over the Earth. It is the least humble way to increase humanity's overall humility. But sometimes it works.

It is noteworthy that the condor, which we have worked so hard to save, is a scavenger, a handmaiden of and symbol for death. The condor is part of a guild of animals who help the dead contribute to the cycle of life, who transfer the energy of the dead back into the living. It was in fact this role that nearly killed them off. When humans wanted "vermin" dead, they poisoned them, and then condors died after eating their strychnine-laced carcasses. When humans wanted "game" dead, they shot them, and the lead shot went on to kill condors that ate the gut piles. In other words, humans were so extremely good at killing, the death we meted out killed *two* nodes of the food web—and maybe more. Who knows what beetles or flies perished after dining on the carcass of a poisoned condor? Alongside captive breeding, condor recovery today focuses on reeling that killing back in, such that hunters or farmers kill just one node on the food web.

If you died in certain parts of the American West today, out in exposed country, you might even be lucky enough to be eaten by the largest bird in North America, to have the energy in your body used to propel those magnificently long black wings through the cloudless desert sky. By capturing the birds and dragging their species back from the brink of extinction, we have preserved that possibility, what ecologists dryly call the species' *function*. And by preventing the extinguishment of *Gymnogyps californianus*, we have preserved *biodiversity*—another dry word for something magical.

A new scientific field was born in 1985, just as the debate over capturing the last wild condors was heating up. Conservation biology was different from its parent discipline, ecology, which simply seeks to describe the workings of the living world. As its name implies, conservation biology has an explicit agenda: saving species. It was built on moral values. One of the field's founders, Michael Soulé, defined the proper objective of conservation biology as "the protection and continuity of entire communities and ecosystems." He identified several "normative postulates," the "value statements that make up the basis of an ethic of appropriate attitudes toward other forms of life." The full list of these postulates is as follows:

- Diversity of organisms is good.
- Ecological complexity is good.
- Evolution is good.
- Biotic diversity has intrinsic value.

Thus today's conservation biologists don't just study what is happening, they make recommendations on what we *should* do, often using the objective-sounding language of science. So for example, a scientific paper in the journal *Conservation Biology* from 2001 says, "To achieve viable wild populations of condors, primary mortality threats must be reduced greatly." This sounds like a scientific statement, and it is. But there's an implied claim about what we ought to do in the sentence as well: We *should* ensure a viable wild population of condors. This claim rests on the value propositions that Soulé spelled out in 1985. I daresay few would disagree with them, all other things being equal.

But "all other things" are not always equal. There are costs to achieving the goal of a wild population of condors—costs in government funding that could have possibly been spent on other environmental or social projects and costs to the condors themselves. Every conservation project has costs of some kind. Some of the projects we will look at in later chapters have very high costs in terms of animal welfare. It is impossible to know whether AC-9 really suffered from losing his freedom for 15 years, and if so how much. (And it's impossible to know whether being captured saved him from a much earlier death by lead poisoning.) But humanity did undeniably use him as a means to the end of saving his species. Whether you believe the ends justify the means in this and other captive breeding cases depends on your values—but it is hard to make such decisions without clarifying our values first.

8

Are Species Valuable?

M any of us have a deeply felt intuition that causing a species to go extinct is wrong. I do. My strong feeling that the diversity of life is valuable and should be preserved is part of who I am as a person. It is in my marrow. Even asking the question "Are species valuable?" feels deeply uncomfortable, slightly heretical. But on this journey, I'm using the tools of philosophy to figure out what humans ought to do in regards to wild animals. So I must be able to clearly and persuasively argue for the value of species if I am going to argue that hurting, killing, or compromising the autonomy of sentient creatures to save species is morally justifiable.

It isn't easy to weigh the value of individual animals against a species. "Species" is an abstract concept, a scientific term that imagines a basket of individual creatures, at a single moment in time, who share certain characteristics. The basket itself is not sentient, cannot suffer or feel pleasure, and is not alive. So why do we value it?

A first-pass approach to this question might be to posit that the moral value of species as such is simply an addition problem—that the value of the species *Gymnogyps californianus* is simply the value of all living California condors, added up. But that can't be right, because then endangered species with very few living members would be *less* valuable than very common species with millions of members. In this accounting, *Rattus rattus* would be vastly more valuable than *Gymnogyps californianus* because there are billions of black rats and about 500 condors in the world.

My sense is that for conservation-minded people, the math actually works the other way. It is almost as if each species has a set value, which is then divided among all the individuals. Say there are 4 billion black rats on Earth. Let's set the value of each species at an arbitrary 100,000 value points. Under this math, each rat would be worth just 0.000025 apiece and each condor would be worth 200. And this valuation matches how we treat them. The rat is despised, shunned, killed without a thought; the condor is cherished, lavished with care and money.

This kind of mental math disgusts animal rights thinkers like Tom Regan. "The inherent value and rights of individuals do not wax or wane depending on how plentiful or rare are the species to which they belong. Beaver are not less valuable because they are more plentiful than bison," he writes.

Martha C. Nussbaum says she does not believe the loss of a species would be a matter of justice "if species were becoming extinct in ways that had no impact on the well-being of individual creatures." She does add that "it may certainly have aesthetic significance, scientific significance, or some other sort of ethical significance." Typically, though, she says, the human actions that drive species toward extinction—transforming their habitats, hunting them, and so on—also hurt individual animals. And so they should be stopped on that basis.

There are, however, plenty of philosophers who do think that species—and the ecosystems they live in—are valuable in and of themselves. The subfield that works on this question is known as environmental ethics. In some ways, the field was founded, posthumously, by a man named Aldo Leopold.

Leopold was born in 1887 and spent much of his career managing wildlife. He was often tasked with killing predators like wolves, which were considered to be vermin. In his essay "Thinking Like a Mountain," he explained the role of wolves in keeping populations of deer down so that they do not overgraze their range. He told the story of his dawning understanding of this relationship as dating from the moment he approached a wolf he had just shot in time to watch "a fierce green fire dying in her eyes." "I was young then, and full of trigger-itch" he wrote. "I thought that because fewer wolves meant more deer, that no wolves would mean hunter's paradise." On the contrary, he found wolf-less landscapes grazed so heavily that anything edible was "defoliated to the height of a saddle horn." Eventually, all the food gone, the numerous deer would run out of food. "In the end," he wrote, "the starved bones of the hoped-for deer herd, dead of its own too much, bleach with the bones of the dead sage, or molder under the high-lined junipers."

It was this ecological view of plants and animals that he wrote about from his rustic "shack" in central Wisconsin in the later years of his life. He was not trained as a philosopher, but he pointed toward a new way of thinking about ethics in his famous 1949 essay "The Land Ethic." Leopold began his argument by noting that in earlier times in human history, morality only applied to people in your own family or band—but not to foreigners or slaves. "When god-like Odysseus returned from the wars in Troy, he hanged all on one rope a dozen slave-girls of his household whom he suspected of misbehavior during his absence," Leopold wrote. This act wasn't seen as wrong. The girls were "property"—not objects of moral concern. But then we changed.

Homo sapiens is a social species. Biological evolution has endowed us with a basic tendency to cooperate with one another. A completely selfish "war of every man against every man"—the "state of nature" described by philosopher Thomas Hobbes as "solitary, poor, nasty, brutish, and short"— never existed. As cultural evolution has built on what biology provided us with, the circle of moral consideration has expanded to cover larger and larger groups of people. Today—in theory, anyway—pretty much everyone agrees that all human lives are valuable. Leopold called for us to enlarge our sense of who or what is part of our "community," and therefore morally valuable even further, to include animals, plants, and the land. In the Western world in the 1940s, this was decidedly a radical idea.

As members of our community, Leopold asserted, plants, animals, and the land have a "right to continued existence, and, at least in spots, their continued existence in a natural state." We shouldn't preserve songbirds because they eat insects that threaten our crops, he argued; we should preserve them because "birds should continue as a matter of biotic right, regardless of the presence or absence of economic advantage to us."

His essay didn't find broad readership until the 1970s, when the culture, it seems, was ready for it. Then, decades after Leopold's death, it not only entered into conversation with the emerging field of environmental ethics, it became a bit of a holy text, especially its principle that "a thing is right when it tends to preserve the integrity, stability and beauty of the biotic community. It is wrong when it tends otherwise."

Leopold's rediscovery came at the same time as pollution, heavy pesticide use, and land clearing began having obvious effects. Polluted rivers were on fire. Smog was so thick in some cities, you could barely see across the street. Rachel Carson's 1962 book *Silent Spring* chronicled the toll pesticides and herbicides were taking on birds and other wild species.

The political response to this moment included many new laws, including, in the United States, the Clean Air Act, the Clean Water Act, and the Endangered Species Act—all of which rested on a basic assumption that people deserved clean air and water, but also that "nature" and wild species deserved protection. Similar laws were passed all over the world. The United Nations Environment Programme was founded in 1972, representing a quasi-global response to the problem.

The philosophical response took shape around the same time, as thinkers began to assemble arguments for *why* "nature" and wild species were worth protecting. Scholars began critiquing the dominant Western beliefs about "nature," identifying an attitude of thoughtless exploitation and strict dualism between "man" and "nature" as problematic. An influential 1967 article argued that these ideas were rooted in the Judeo-Christian idea that God gave humankind dominion over all other species.

Traditional ethical theories up to that time primarily focused on human beings as the center of the moral world. A new generation of environmental ethicists saw that this anthropocentric focus wasn't a good fit for thinking about the environment. Sure, you could argue that it was wrong to cut down a forest because humans enjoyed hiking and camping in it, but then again, humans also enjoy living in houses made of wood and reading books made of wood pulp. And what of places or species with few or no human fans? Shouldn't boggy marshes and nondescript insects have the same claim to protection as the Grand Canyon and the Bengal tiger? Maybe Leopold was right. Maybe "land" can have a right to exist; maybe "biotic communities" are valuable in and of themselves, no matter what people use them for.

Philosophers make a distinction between "instrumental value" and "final value." Instrumental value is the value something has because someone can use it to get something else they want. Money is valuable because you can use it to buy food, shelter, and healthcare. We often talk about the instrumental value of other species in terms of "natural resources" like timber, game animals to hunt or fish, and pollination services provided by free-ranging insects. Note that species don't have to be useful to *humans* to be instrumentally valuable. Alder trees might be valuable to a human as a source of firewood or a protective screen from the wind. They might also be valuable to a beaver as food. In both cases, we can say that alder are instrumentally valuable.

However, many of us feel in our hearts that species of plants, animals, and fungi are valuable even if there are no people (or beavers, or other sentient creatures) around. These species are valuable inherently, just for being themselves. This is "intrinsic" or "final value."

When animal rights thinkers say that you cannot use a sentient being as a means to an end, the basis for that argument is that sentient beings have final value. Conscious selves are valuable in and of themselves. The question here, though, is whether a *species* has final value.

There's another key distinction within the category of final value, and that's between objective and subjective value. To say that something is objectively valuable is to say that it just *is* valuable. Its value is a true fact about the world. Even if there are no human or non-humans around that value the thing, it is still valuable. As philosopher Ronald Sandler says, "Objective value is discovered by valuers; it is not created by them." In

1973, a philosopher named Richard Routley—who later changed his name to Richard Sylvan—proposed a thought experiment. Imagine that the last person on Earth kills every living thing on the planet—all plants and animals. The animals are killed painlessly. Has this person done something wrong? For many, the idea of the last human on Earth destroying every living thing on the planet before he or she dies isn't just clearly wrong, it is also deeply sad and even repulsive, even though there would be no sentient beings around to mourn the loss. This thought experiment hints that species and ecosystems are objectively valuable.

To say that something is *subjectively* valuable is to say it is valuable *because* someone values it. But we are talking about final value here, remember, so these valuers aren't valuing it for its usefulness. They are valuing it for what it is. As Sandler says, "[T]he valuing might be for what the entity represents, for what it embodies, for its rarity, for what it expresses, or for its beauty. But in each case, the valuing is for noninstrumental reasons, not for what the entity can bring about." It is the act of valuing that creates the value. If all the valuers go away, the entity loses its value.

So, to sum up, if you love the species *Oncorhynchus tshawytscha* (Chinook salmon) because you like to eat it, then you value it instrumentally. If you value it for cultural or religious reasons, as many people do in the Pacific Northwest, or if you value the beauty of their silver-and-crimson backs visible through the clear water of a salmon stream—or the power of the story of their return from the Pacific Ocean to spawn, then you may be conferring upon it subjective final value. But if all the people and all the bears and all the other species that value salmon went away—poof—and the species *Oncorhynchus tshawytscha* was still valuable—*separately from all the individual fish that make it up*—then the species has objective final value.

For a long time, as I read about all these flavors of value, I gravitated toward the notion of "subjective final value." It seemed to capture the almost mystical worth that I felt the non-human world carried, but identified a clear and defensible location for that value—in the hearts (or, to be more literal, brains) of people. But there's a persuasive argument against this idea, one Richard Routley articulated all the way back in 1973. What if people don't value non-human species the way *I* think they should? What if other people just don't care about the non-human world? "Whether the blue whale survives should not have to depend on what humans know or what they see on television," he wrote.

Humans value the wrong things all the time. Racists love their own race and not people of other races. And they are wrong. Some humans love oil and gas profits more than coral reefs. I think they are wrong too. Not everyone values the same things or to the same extent. Many people find large swathes of the outdoors too hot, too cold, kind of creepy, or just plain old boring. People's beliefs about what is valuable may be based on misunderstandings or incorrect information. And, as Sandler points out, people's values are also often inconsistent: "Someone might be outraged by the practice of euthanizing unwanted animals, but delight in having bacon with every meal." Is it really a good idea to build an ethical system on the variable, erratic, poorly-thought-out judgements of people?

"For these reasons," Sandler writes, "many environmental ethicists believe that the justification for such things as wilderness protection, species preservation, and compassion toward animals are only fully secure if these entities—ecological systems, species, and animals—have objective value or value that exists independent of anyone's actual evaluative attitudes." In other words, many ethicists *want* these entities to possess objective final value because that is the best way to protect them. They want this value to be as universally acknowledged as the worth of human lives.

I share this desire. But I am less certain that environmental ethicists can prove that species and ecosystems have objective final value. Let's dig into species first. Individual animals can feel pleasure and pain; things can go better or worse for them. No one claims that a species can feel pleasure or pain, since it is a category, not an organism with a sense of self. But some say that things *can* go better or worse for species. And that sounds reasonable at first. When species are dwindling in numbers and heading toward extinction, it seems like things are "going badly" for the species, even if each individual in it is dying painlessly of old age after a rich and fulfilling life.

But Sandler points out a flaw with this analogy between a species and an individual. An individual living thing, sentient or not, tries to stay alive and reproduce. You can quibble with the word "try" for a plant, but given what we know about how they react to insect attack by releasing toxic chemicals into their leaves; how they grow toward the light; how they respond to their environment, always in a way that tends toward life, growth, and reproduction, I think it is appropriate. Plants have goals. Thus, when they grow big and tall and spread their canopy wide to suck in the sunlight and then disperse lots of seeds, we can say that things are going well for them. When they die of bark beetle infestation, we can say that things went badly for them.

Species, however, are not goal-directed in this way. They do not try to persist or flourish. It may seem like they do, because we want them to persist and because each individual that the species comprises tries to survive, but the species itself does not take any action to prevent its own extinction. And the reason it doesn't, while even nonsentient living individuals do, is that evolution does not "select" the fittest species. It selects the fittest individuals.

The individual living things alive today are here because their ancestors survived and reproduced. And thus each one has inherited many ways of surviving and reproducing. Their goal-directed nature is a product of the brute force of evolution. Individuals that don't try to live and reproduce don't have babies. Those that do, do, and pass on their "trying" genes. Sometimes selection can act on tightly coordinated groups, like ant colonies, when the whole colony survives or dies together. But "species are too diffuse and their individual members too uncoordinated and independent from each other for them to constitute an entity on which selection might operate," as Sandler says. Species don't try. They have no goals. If they have no goals, they can't be helped or hindered.

There is another argument for why species might have objective final value, an argument laid out by the environmental philosopher Holmes

Rolston III. In his account, the value of a species derives from the long evolutionary process that created it—and from the potential it contains to continue to evolve in the future. To cause an extinction is to "shut down a story of many millennia and leave no future possibilities." I agree that this is a tragedy in the sense that I agree that I don't like to see that story shut down and that future cut off. But does its wrongness emerge from the fact that I, Rolston, and countless others don't like it—which would be subjective final value—or is it objectively bad?

Rolston says, without hesitation, that humans causing an extinction is objectively bad. He asserts that this is a true fact about the universe. Sandler says that just asserting that a long evolutionary story is valuable doesn't make it so. He admits that many people feel that millennia of evolution makes biodiversity valuable. But that doesn't prove anything except that people are fond of biodiversity. "In cases like this," he writes, "the absence of reasons to believe is strong evidence not to believe."

A large part of me wants to sign up for team Rolston III and just assert that millions of years of evolution being snuffed out in an instant *has* to be wrong. But I have doubts. Evolution is not a "good" process. It is undirected by any intelligence, loving or otherwise. It is completely amoral. All individuals are slightly different. Some have babies and some don't. Babies inherit the traits of their parents, who managed to reproduce. Repeat that for millions of years, and you get 10 billion species. You get the California condor and the aspen and the beaver and the elephant and the tiger and the ʻakikiki and ʻakekeʻe. I *delight* in these different forms of life, but the process that created them is not good. It is just time and sex and death and mutation and chance.

———

Species get a lot of attention in conservation, but they aren't the only way to look at "biodiversity." Many conservationists are oriented around species and tend to proceed as if every recognized species is equally valuable. But some rank the value of a species based on its genetic distinctiveness from other forms of life. The EDGE of Existence program identifies and prioritizes species that are "Evolutionarily Distinct and Globally Endangered"—giving them the acronym EDGE. Their philosophy is that species that have few or no close genetic relatives are more valuable. These species have been evolving independently for a long time, and so "represent a larger amount of unique evolution." More of their genes are exclusive, found in no other species in the world. Here, the species are really conceptualized as baskets of genes, and the diversity of *genes* is the conservation goal. Instead of the usual tigers and pandas and condors, they focus on animals like the Hispaniolan solenodon, a tiny long-nosed creature that is "one of the last representatives of an ancient lineage of shrew-like mammals that lived with dinosaurs from 76 million years ago" and the largetooth sawfish, a tropical relative of sharks with a nose like a chain saw.

The EDGE approach takes seriously the idea that "biodiversity" is more than just counting up the number of species in a particular place. Nine times out of ten, when conservationists talk about biodiversity, it is being used as shorthand for "number of species," but in textbooks and other educational contexts, the concept is usually defined more grandly. The American Museum of Natural History defines biodiversity as "the variety of life on Earth at all its levels, from genes to ecosystems, and can encompass the evolutionary, ecological, and cultural processes that sustain life." (This particular definition is even more than usually expansive in that it includes "cultural processes"—a welcome hint that human beings can fit inside this concept.)

What these broader definitions suggest is that "many species existing" is not enough. We don't want "biodiversity" preserved only as genes in a -80° freezer or individuals living in captivity. We want lots of genes made concrete in an abundance of individuals, and we want those animals in ecosystems, interacting with one another in their characteristic ways. This is what really makes the conservationist or the outdoors-person happy. But why? What is so great about a bunch of rocks and water and dirt and plants and animals and fungi all living and dying and eating and being eaten and singing and mating and reproducing and competing and helping each other and evolving in response to one another?

Ecosystems, like species, can be valued instrumentally, for the "services" they provide humanity. Ecosystems produce the oxygen we breathe, filter the water we drink, act as living shields between hurricanes and human communities, host plants and animals we harvest for food, fiber, and medicine, and more. Forests, mountains, and beaches provide us places to recreate and relax—another instrumental value. And of course ecosystems have subjective final value in abundance. People love forests, deserts, marshes, and coral reefs with a passion. There's something about the *interactions* between the individuals that seems to add up to something greater than the sum of its parts. And many argue that ecosystems, like species, also have objective final value. This is what Leopold was arguing for, in essence.

Leopold chose to couch this value in terms of "integrity" and "stability" (as well as beauty), suggesting that ecosystems don't really change unless people change them. This framework makes a certain amount of sense in a world where among the biggest threats to the non-human world are bulldozers and pavers and chain saws—a world where people are physically dismantling complex ecosystems to extract their profitable bits or transform the place into something else. But as we saw in Chapter Four, the "ecosystem integrity" that conservationists seek to preserve is based on one moment in human history, often the moment of colonization. It represents a frozen instant in the long unfolding of a place. And even in the absence of human intervention, ecosystems are not actually stable or unchanging.

When the climate changes—whether we're talking about ice ages thousands of years ago or anthropogenic climate change now—species react in part by moving on the landscape. Ecosystems do not move as units, though; each species has had its own trajectory over the millennia, evolving and shifting ranges, sometimes sharing systems with other species for a time, then parting ways. As ecologist Henry Gleason said as far back as 1926, "Every species of plant is a law unto itself." No ecosystem, defined by its composition of plants, is older than about 12,000 years. And each ecosystem varies in space across its geographic spread.

Thus ecosystems are not really like organisms. They are not wholes. Their parts interact, but those interactions are not eternal. When two species interact consistently for many millennia, they may adapt to one another so specifically that we call that relationship a "coevolution." The classic example here is *Xanthopan morganii praedicta*, sometimes called Darwin's hawk moth. In 1862, Charles Darwin studied the beautiful Star of Bethlehem orchid from Madagascar, which features a very long tube with nectar and pollen at the end. Darwin predicted that a moth with a proboscis of the same extravagant length would be found, and sure enough, 41 years later, a plump brown moth was discovered, with a foot-long proboscis.

Such relationships are found across the world between predators and their prey, pollinators and the plants they pollinate, parasites and their hosts, and more. They are a hallmark of lengthy associations between species, so they are less likely to be found in ecosystems that have come together recently. This lack of deeply coevolved relationships in rapidly changing ecosystems is one argument some conservationists make for why the "integrity" of ecosystems is important. Coevolution is undeniably cool, so I understand their thinking. On the other hand, new coevolutionary relationships can and are being created all the time. Give these newer ecosystems some time, and they will eventually develop the same impressive coevolutionary relationships. Unfortunately, it may take quite a bit longer than a human lifespan.

One of my favorite studies from the last few years is a look at the introduction of North American black cherry trees into Europe. Menno Schilthuizen at the Naturalis Biodiversity Center in Leiden, the Netherlands, and his colleagues looked at leaves of the cherry in herbarium collections going back 170 years. At first, all these leaves looked nearly perfect. European insects that eat leaves didn't have much experience with the tree; they were not attracted to it and they were put off by the cyanogenic defense compounds it made. But over the years, the leaves got more and more ragged and nibbled as European insects adapted to the new food source. Black cherry was spreading rapidly and its leaves represented a big pool of energy. Of course at least some species would adapt to take advantage. Eventually, the leaves were as be-munched as those of the native cherry.

This process helps explain the boom-bust pattern seen in many introduced species. At first, no one in the new environment knows how to eat it, and so it thrives and reproduces prodigiously, often causing considerable worry as humans see it spread and imagine how dominant it might become if it keeps multiplying at such a dizzying rate. But the more it spreads, the bigger potential food source it represents, increasing the incentives for other species to learn how to eat it. Once that happens, the rate of reproduction takes a serious hit. Then it must adapt too. Schilthuizen has found hints that the black cherry has adapted its defense compounds to better repel European herbivorous insects. As other species adapt to use it as a resource, and as it adapts back, it becomes woven into the food web, naturalized to the point where it is not a monstrous threat. Thus, well-meaning efforts to kill or control new species like the black cherry may accidentally do the opposite—by squelching the boom, they slow down the process of naturalization, slow down the formation of those delightful coevolutionary relationships. Indeed, in time, the European black cherry might diverge from its parents in North America and become a new species altogether, thus increasing biodiversity!

The ecologist Chris D. Thomas at the University of York in the United Kingdom expects that introduced species will adapt to their new homes by eventually diverging into new species—and that these new species will come to be a major source of new biodiversity. "Come back in a million years and we might be looking at several million new species whose existence can be attributed to humans."

Ecosystems typically change at a pretty rapid clip from a geological perspective, but without human intervention, these shifts are often too slow for us to notice. We only live a century at most apiece. And so, conservationists have tended to acknowledge that ecosystems change in their theory but treat ecosystems as basically static in their practice. "Ecological restoration" was defined in the 1990s as "the process of intentionally altering a site to establish a defined, indigenous, historic ecosystem. The goal of this process is to emulate the structure, function, diversity and dynamics of the specified ecosystem."

As the field has matured and changed, especially as it has reckoned with climate change, the role of history in defining goals is being more openly questioned, which I think is all to the good. Trying to manage for "integrity"

or historical states can lead to perverse outcomes, especially when we intervene to turn back time even as evolution surges ahead. We can find ourselves
fighting against processes—from hybridization of related species to range
shifts to the evolution of new ecological relationships—that would create
complex, resilient ecologies suited to the changing world. It almost seems
like a failure of humility.

Environmental ethics aren't easy. After all my study, I am still not sure I
can come to final conclusions about what is valuable. Adult Condor 9 had
final value as a sentient creature, as do all living birds. I feel confident there.
Gymnogyps californianus is valuable as a species *to humans*, who find it
majestic and inspiring and fascinating (and do not want to feel guilty and
bad if it goes extinct). Humans also value the species as a unit of biodiversity. Were it to go extinct, biodiversity would be lessened by one species, and
any further evolution of the lineage would be cut off.

The ecosystem in which the condor lives is also more valuable *to humans*
when the condor is present, because it has more species and more interactions between species.

Some also think the ecosystem is more valuable with the condor in it
because the condor was there in the past. And I think that valuing previous
ecosystem states is fine as long as we all agree that this preference is inherently cultural, not scientific. For example, the Yurok Tribe's interest in
bringing back prey-go-neesh is explicitly cultural, and isn't appealing to
some quasi-scientific notion of "ecosystem integrity." Managing Yellowstone
to look like it did in 1872 because so many Americans want it to stay that
way is also acceptable, as long as we are being honest about the fact that this
date emerges from history, not ecology.

There are lots of compelling reasons why condor recovery is important
to humans. But can I confidently assert that *Gymnogyps californianus* as a
species, separate from the individual birds that comprise it, does it have
objective final value? Do I agree that "biotic diversity has intrinsic value"?

I don't know.

Just writing that sentence frankly terrifies me. I want to go dunk my head
in a snow-fed creek and wash away my doubts. But I can't. In a way I cannot
explain rationally, I think that the diversity of life and the complexity of

ecosystems *matter*, even if there is no one around to care. I think that there's something precious in what we call "nature," in the flow of energy, in the will to survive, in the way a lupine leaf holds a perfect sphere of rain. But I cannot present overwhelming arguments that this is true. I can only passionately assert it, like Rolston III.

To reiterate: I do not think that preciousness resides in stable, closed entities—species or ecosystems—that are like hard jewels. Species and ecosystems should not be frozen in time, prevented from changing or mixing. That would arguably make them increasingly vulnerable to collapse. I *like* the way they surprise us with their own solutions to climate change and other problems. I value exactly that dynamism, that unpredictable complexity, that strange beauty. But I cannot prove that it has objective final value. Perhaps the virtue of humility also applies here—humility about what we know and what we can prove.

Remaining agnostic about the objective final value of species or ecosystems does not mean that we should not attempt to prevent extinctions. The value that a species like *Gymnogyps californianus* has to us humans is immense. It certainly justifies the tens of millions we've spent on saving them. And I am inclined to say that it also justifies any suffering and loss of autonomy experienced by the captured birds, especially since the levels of suffering seem quite low in this case—lower than what awaited them in a "wild" lousy with lead shot.

All other things being equal, the combined instrumental and subjective final value of any species is high and warrants extraordinary efforts to preserve it. But without being able to definitively prove their objective final value, things get a little trickier when saving a species means causing harm or significant suffering to sentient animals. Because now we cannot say we are causing that harm to preserve a timeless universal good. All we can say with confidence is that we are doing it for ourselves.

9

Feeding Polar Bears

O f all the actions humans take to help wild animals, perhaps the most widespread is feeding birds. Forty percent of Americans feed wild birds, and in the United Kingdom, the rate is an impressive 75 percent. We feed birds so they will come close and hold still for a moment so we can get a look at them, of course, but we also do it to benefit them, especially in winter, when we have the vague impression that they are hungry and cold. Feeding birds can increase their survival and boost the number of chicks that fledge. And in a world where they must deal with our windows, our wind turbines, and our voracious pet cats, it seems like such feeding could help make up for the many ways we reduce their chances of survival.

On the other hand, feeders can become crowded hot spots for disease transmission. Parasites can spread as birds press close. And feeding can arguably reduce "naturalness"—depending on how you define the term. Some Eurasian blackcaps have given up migrating south for the winter because of the availability of feeders. In the case of the blackcaps, the birds that have quit migrating are actually diverging into an entirely new species from their kin that have kept up the annual trek to the south. The stay-at-home birds are evolving longer beaks, better for the seeds available at the feeders than the fruits the migrating birds specialize in.

If feeding birds is largely seen as harmless or even good for wild birds, feeding mammals is generally frowned upon. Those who put out corn for deer or cat food for raccoons are chastised for habituating animals to the human world, a habituation that can lead to their death—whether by car accident or by violence when their presence in the human sphere terrifies or threatens a human or their household. In Klamath Falls, where I live, someone fed deer in a neighborhood at the edge of town. The deer became so numerous on this block that they attracted a mountain lion. The cougar was killed by state wildlife officials once it became clear that he intended to stay permanently. In that case, one man's love for deer cost a big cat his life.

My family and I were on a road trip once and stayed a few nights in Colorado at a house owned by a friend who was out of town. It turned out that their neighbors had been feeding a fox, who had gotten in the habit of visiting the adjoining suburban backyards and sometimes laying in the shade on their deck. The fox never let anyone pet it, but my friend told me that if her husband kept very still, the fox would sometimes sniff at his foot. This fox seemed interested in us and made as if to approach. I was so militantly committed to "keeping wild animals wild" at the time that I forced my family to scurry back inside when the flame-colored animal came near. My three-year-old daughter lingered by the glass door to the deck, instinctively drawn to the fox, desperate to talk to it, touch it, befriend it.

We badly want to connect with wild animals; but so often, it doesn't turn out well for the animal, even if they too seek the connection. An orca known as Luna—or Tsux'iit to the Mowachaht/Muchalaht First Nation—was two years old when he somehow got separated from his family. In 2001, he was found living alone in Nootka Sound, in British Columbia, calling out for kin, but getting no reply. So he began approaching boats and docks, where people could tell he was lonely. So they interacted with him. "How can you *not* touch the whale, when he comes over there?" asked Paul Laviolette, a logger from Gold River, who had been approached by Luna. "It was a beautiful feeling, communicating with that animal like that." In a documentary film about Luna, countless people recall gazing into his eyes and feeling a deep bond with him. "It often seems that there's a wall between us humans and wild beings, built of fear, and respect," said filmmaker Michael Parfit. "Luna was breaking it down." But all those boats came equipped with propellers, and he risked being injured in every encounter. In 2006, Luna got too close to a tugboat propeller, and died.

So we make rules against feeding wildlife, against touching wildlife and making friends. But could there be a moral obligation to feed animals in cases where they are starving because of us? After extreme wildfires linked to climate change tore through parts of Australia in 2019 and 2020, experts said it was okay to put out water for koalas, brush-tailed possums, and other suddenly homeless wildlife. But feeding was felt to be a step too far. "Supplementary feeding isn't advised unless habitat and sources of food have been completely destroyed, and is only appropriate as a short-term emergency intervention until natural resources recover," the biologists wrote. "But leave it up to the experts and government agencies, which provide nutritionally suitable, specially developed and monitored food in extreme cases."

After a wildfire in California, officials at the Santa Monica Mountains National Recreation Area were even tougher. "Want to help wildlife after the fire? Do not approach wildlife & please DO NOT LEAVE OUT WATER OR FOOD, as it may habituate them & have other unintended consequences," they tweeted. "Wildlife are highly resourceful. It is against the law to feed or disturb wildlife on [National Park Service] lands!"

Polar bears have become a symbol of climate change's effects on the non-human world. In 2017, *National Geographic* posted a video of a starving polar bear, its long body gaunt and fur patchy. The magazine estimates the clip was watched 2.5 billion times. Some viewers were angry that the videographers had not somehow intervened to save the bear. "Of course, that crossed my mind," photographer Paul Nicklen said. "But it's not like I walk around with a tranquilizer gun or 400 pounds of seal meat."

The IUCN lists polar bears as "vulnerable" and says that because the bears eat seals, which they can only hunt when there's a thick layer of ice on the Arctic Sea, "climate warming poses the single most important threat to the long-term persistence" of the species. A study of nine female polar bears, using collars equipped with video cameras to record their behavior over about 10 days in the springtime, showed that they must eat "either one adult ringed seal, three subadult ringed seals, or 19 newborn ringed seal pups every 10 to 12 days" to keep from losing weight and sliding toward starvation. Bears scavenge beached whales and dead seals, and occasionally stalk prey, but their go-to hunting behavior is to sit and wait by one of the breathing holes the seals maintain in the ice. When a seal pops up for a lungful of air, they bash it in the head with their mighty paws. But the sea ice is shrinking by 14 percent every decade, thanks to the ever-thicker blanket of carbon dioxide and other greenhouse gasses we humans are constructing in our shared atmosphere, and the winter ice doesn't last as long as it used to.

However, the whole truth is, as always, more complex than this simple narrative. Experts divide polar bears into 19 populations, and each has its own story. Some, like the northern and southern Beaufort Sea populations, are categorized by IUCN specialists as likely declining owing to a shrinkage in the total area of their prime hunting grounds over time. But 72 Inuvialuit elders—who know polar bears well, since the Inuvialuit have been hunting them for longer than living memory—agreed in a 2015 report that "the general health and abundance of polar bears in the Beaufort Sea region remains generally stable. . . . However, there appear to be fewer really big bears and no bears are as fat as they were prior to the mid-1980s." Some polar bears may be changing their behavior to cope with the loss of sea ice. Certainly, more polar bears are looking for food in and around human communities. They congregate at dumps, and hunters are increasingly finding plastic in their stomachs.

In other subpopulations numbers are actually going up. And many of the 19 populations are considered "data deficient"—in other words, researchers don't know if polar bears there are declining, stable, or even increasing. The IUCN estimates that there are somewhere around 26,000 polar bears in the world, in Norway, Alaska, Greenland, Russia, and Canada, and there's been about that many for decades. So are the bears in danger or not? The polar bear is not in imminent danger of extinction, experts agree. But declines in the sea ice are predicted to reduce the number of bears by about 30 percent by 2050.

That statistic may sound comforting—the polar bears will live on! But behind that 30 percent decline are starving bears, dying cubs, stress and sadness and pain for thousands of bears. Clare Palmer, a philosopher at Texas A&M in College Station, uses the polar bear as a case study for thinking about assisting wild animals. Should we feed bears whose core territories are seeing substantial loss of sea ice due to global warming?

Palmer acknowledges that feeding wild polar bears goes against a strong feeling many of us have that wild animals should be left alone. Palmer has named this feeling "the laissez-faire intuition." When someone neglects their horses, letting them sicken or starve, we are appalled and demand punishment. But when wild deer sicken or starve in a harsh winter, most of us feel no need to intervene. As my ten-year-old daughter says, "It is not our business."

The laissez-faire intuition makes a lot of sense to most of us. We talk about "letting nature take its course" and "leaving no trace." Drawing a line between us and the wild, then staying firmly on our side of that line, is neat, tidy, and sounds doable.

Palmer believes that this intuition tells us something true. We do have different obligations to our pets and our domestic animals than we do to wild animals. Humanity is a partial creator of domestic animals, and because of this, we bear more responsibility for their welfare. Wild animals aren't our creations—at least not our intentional creations. But if humans harm wild animals by, for example, knocking down the woods they live in to build a new housing development, then this creates new obligations to those animals, Palmer says.

As we have seen, though, our entanglement with wild animals is knottier now than it was for our distant ancestors. Through climate change, our massive transformation of terrestrial ecosystems, and other thoroughgoing influences, we have determined the course of many wild animals' lives nearly as much as we've determined the courses of our farm animals or pets' lives. As animal welfare expert Lotta Berg from the Swedish University of

Agricultural Sciences in Skara told me, "They say we should leave them alone but the problem is that we're not. Not intervening is also an animal welfare decision."

In the polar bear case, Palmer argues that humans are morally responsible for the consequences of climate change. She admits that "the details of how this moral responsibility is distributed over human nations, populations and individuals over time and space, are contested, as is determining who has responsibilities to do what as a result," but speaking broadly, she says, humans caused climate change and have some kind of collective responsibility to those who are harmed by it.

The notion of collective responsibility is a key part of many arguments for environmental action. Often, the specific people who caused a given problem are diffuse, unknowable, or dead. Except in cases where particular governments or corporate entities caused major environmental harms, it is usually not feasible to round up the exact people who are to blame and make them set things right. In some cases, like the decline of whales, the bulk of the harms happened a generation or more ago. Few people alive today have personally killed a whale. But many of us share a sense that "we" still have some kind of responsibility to protect and cherish the whales we have today.

Collective responsibility can be tricky. One way out of its many difficulties is to shift our focus from who *caused* the harm to who *benefits* from the harm right now. Saying that "humans" are causing climate change is, roughly speaking, accurate; few humans truly emit zero greenhouse gases in the course of their lives, though many humans contribute far fewer molecules of carbon dioxide than I do, as a middle-income American. But it seems obscene to say that, because a poor rice-farming family in Bangladesh uses a gas powered tractor, they are partly responsible for climate change and need to do something to make good—while climate-change-linked floods kill their crops as they stand in their fields. If we instead assign responsibility to those who *benefit* from climate change, our farming family is off the hook. Instead, the biggest chunks of responsibility go to people like fossil fuel company executives (or to the companies themselves, under the legal fiction of corporate personhood) who make very large profits from selling the stuff that produces greenhouse gas emissions. Smaller but not negligible fractions of responsibility go to families like mine, who have benefitted in many ways from cheap energy, including enjoying recreational travel, and whose modest retirement funds might be making money on fossil fuel company stocks found in big aggregated funds.

Those who emit because they are trapped within a socio-economic system built around fossil fuels but are not *profiting* off climate change are, under this interpretation of responsibility, victims, not villains. So in a scenario investigating whether "we" should take some costly action to stop or reverse global warming, I think the "we" should be understood as the group of people who benefit from climate change.

Let's return to Palmer's argument. What does she think "we" should do about the bears? Firstly, we should stop cooking the Earth, so as to turn off the harms. But that won't fix the sea ice problem right away. Even if we straighten up and stop emitting *tomorrow*, some more warming is baked into the system and it will be a long time before the trend is reversed.

Palmer says that in the meantime, there are good reasons not to just let the bears starve. Her favorite reason is that doing so would be unjust to the bears. Polar bears are sentient creatures that have been harmed by climate change, and, they have not shared in any of the benefits. Glenn Kellow, the CEO of Peabody Energy, the world's largest private coal company, took home nearly 8 million dollars in 2019, and the polar bears are starving to death partially as a result. Even a child could tell you that's unfair.

Another way to put the justice argument is as a matter of rights. Remember that if rights are the things that everyone is entitled to, justice is the practice of making sure everyone gets those things. Here the rights being infringed upon are the polar bears' rights to access food and maintain health and life. And Palmer says that even if you don't like the justice framing. There are still good reasons for helping the bears based on utilitarian calculations about suffering or their preferences not to starve.

No matter how you justify it, though, the fact that there are good reasons to help doesn't mean that there aren't also good reasons *not* to help. Some are quite practical. How and what would we feed the bears? Feeding stations would have to be set up well away from human communities where the bears would almost certainly become a dangerous nuisance. And would we feed them their typical diet of seal meat? Because if so, we will need to employ some seal hunters and—wait a minute. Aren't seals sentient creatures too?

This brings up a problem that is bigger than this particular case: helping predators satisfy their interests, feel pleasure, or express their innate capabilities—however you want to put it—seems to necessarily include allowing them or even helping them kill other sentient beings. Back when polar bears and other predators were just other animals going about their lives, this wasn't necessarily our business. Now that so many predators are suffering because of human hunting, climate change, habitat loss, and other factors that we bear some kind of collective responsibility for, we now have enhanced responsibility to care for them. But caring for them means enabling them to hunt, kill, and eat. For example, the widely-praised captive breeding program that saved the black-footed ferret required conservationists to kill thousands of prairie dogs, golden hamsters, and other small mammals to feed them. This is indeed a moral dilemma.

In the polar bear case, Palmer says, you can work around this problem. Polar bears could instead be fed byproducts from slaughterhouses, the fat and bone meal that humans don't eat, which is normally used for pet food and fertilizer. If humans are going to keep eating meat, she says, then using the waste from this process seems less ethically problematic than killing seals for the bears to eat. Hers is a pragmatic workaround that doesn't tackle the ethical problems of the meat industry itself. One could also imagine a solution involving creating plant-based food that would be both delicious and nutritionally appropriate for polar bears. But the more general problem of predation is going to pop up again and again.

Palmer adds that the bears we save from starvation during the ice-free period might well turn around and head out on the ice as soon as it forms and eat a bunch of seals. Will those seal deaths then be on our heads? Utilitarians will want to know how to weigh the suffering of the polar bears starving against the suffering of the seals they eat later. If each bear kills and eats more than one seal, they might suggest that painlessly euthanizing the bears would be ethically superior. Justice theorists might argue that our supplemental feeding has just held the number of bears steady rather than allowing it to crash, so the level of predation risk hasn't changed for the seals and thus we haven't wronged them.

But ecology is extremely complex. Predicting the ripple effects of the local extinction of polar bears in terms of the pain and pleasure of all the sentient beings in their former sphere of influence is next to impossible with our current levels of understanding.

One might also object to feeding the bears because it is "unnatural." As I've said, I don't think "naturalness" is valuable. But I do think the autonomy of individuals is. So I share the concern that the feeding program could "change bears' behavior in ways that reduce their welfare by limiting agency or increasing suffering down the road, if the fed bears were no longer able, or willing, to hunt for themselves," as Palmer says. You could imagine the length of the supplemental feeding program growing longer and longer each year, until one day the managers realize that they are feeding the bears year-round and that they now have hundreds or maybe even thousands of massive obligate carnivores needing 12,325 calories a day depending on them. Such bears might fail to teach their babies how to hunt; such babies might grow up to be hopelessly inept on their own; we might even see evolution select for the ability to tolerate a high density of polar bears that thrive on pork fat.

Such dependency might not bother the bears particularly, though we can't know what an unfulfilled longing to hunt or wander great distances alone might feel like for a polar bear. But it would definitely make them vulnerable. If our hypothetical Project Polar Bear got its funding cut and the pork fat or Vegan Polar Bear Chow vanished overnight, those future bears might suffer a lot.

Even if the program were securely funded and able to operate in perpetuity, year-round bear feeding gives me a feeling not unlike the one voiced by contributors to *The Condor Question*. It seems, frankly, undignified for the bears.

One way to think about our relationships with wild animals is to think of their communities as other nations with a right to sovereignty, according to philosophers Sue Donaldson and Will Kymlicka, both affiliated with Queen's University in Ontario. "This means that if and when we humans visit their territory, we do so not in the role of stewards and managers, but as visitors to foreign lands," they write. "[A] sovereign community has the right to be free both from colonization, invasion, and exploitation on the one hand, and also from external paternalistic management on the other." Looked at through this lens, permanent dependency on humans would be like the community of polar bears being taken over by another country. It could even be seen as a kind of colonization.

Wild animals flourish in communities made up of their own and other species. When humans want to develop those lands, they often argue that they are "unused" or "empty." Donaldson and Kymlicka cite environmental scholar Jennifer Wolch, who explicitly compares this to the "terra nullius" doctrine used to ignore Indigenous claims to lands colonizers wanted. Donaldson and Kymlicka say that respecting the sovereignty of wild animals doesn't mean never entering their territory or intervening in their affairs, though. "Animals have evolved to survive under these conditions, and are competent" to withstand them, they write. But while "respect for sovereignty . . . rules out systematic intervention to end predation or natural food cycles," thoughtful aid that "allows a community to get back on its own feet" could be ethically acceptable. Interventions like helping animals harmed by pollution or climate change or even a deadly virus would be akin to "foreign aid" to nations suffering an overwhelming catastrophe. So if handled carefully, a feeding program might qualify as acceptable. After all, sovereignty isn't much use to you when you're dead.

Freya Mathews, a philosopher at La Trobe University in Melbourne, Australia, says that "where there is no way to restore habitat, we should give wild animals the opportunity to adapt to a human-mediated environment and let them choose for themselves"—calling such a decision "the ultimate exercise of wild sovereignty." So for Mathews, it is right to offer polar bears food if their seal hunting grounds melt. If they take it, it could mark the beginning of a perpetual or at least long-lasting covenant between our species.

Nationhood is a compelling metaphor, a way to make sense of the laissez-faire intuition even if, like me, you aren't sold on the value of "wildness" understood as "lack of human influence." Even Tom Regan talks about wild animals as "other nations." Sovereignty is about physical autonomy, not genetic or ecological purity.

There are reasons beyond compassion for members of animal nations to consider helping starving polar bears. Palmer writes that "an Arctic landscape that retains polar bears may be regarded as more natural than an Arctic landscape that lacks them." But given everything we know about the word "natural," I think rather it is more useful to say that an Arctic landscape that retains polar bears *is more desired by humans* than an Arctic landscape that lacks them. The human love for polar bears should count for something in this discussion. The Inuvialuit want to hunt them and to watch them. Lots of people want to see them in their Arctic habitat—or perhaps they don't actually want to see them, but they want to know that they are up there, ivory white on pure white, their great necks outstretched and their little cubs walking behind them. I think this kind of subjective final value should be considered here too.

I believe, with Palmer, that our massive influence on non-human animals living today creates an enhanced responsibility toward them. Collectively, we've made the world that all these sentient creatures inhabit; we have at least some duty to them. What that duty entails continues to bedevil me. How do we balance our enhanced responsibility to the non-human world with respecting the autonomy or sovereignty of non-humans while striving to remain humble? If we do too little, the world remains unjust and non-humans suffer; if we do too much, we arrogantly dominate and oppress. And here's what worries me the most: what if there is no sweet spot?

"Unfortunately, this situation presents a choice between only bad options," Palmer concludes. She decides that more information about the outcomes of feeding is needed and proposes a feeding trial, if such a trial is acceptable to the Indigenous communities in the area.

This last point is important. We've been discussing the idea that humans can have obligations to individual animals harmed by climate change, that helping them may be a matter of justice. Of course, humans can also have obligations to individual *humans* harmed by climate change—and Native communities in the Arctic and subarctic are seeing some of the biggest, fastest changes to their ecosystems, threatening their economic and cultural security. The "climate justice" movement explicitly seeks to wrestle concern about climate change away from just coral reefs and polar bears to refocus on the many, many humans who are being and will be harmed. As we think through complex environmental and animal-welfare problems, we must always consider the human beings involved as well—and there are nearly always humans involved in these cases. Humans are a part of most places, and they deserve both justice and care.

Feeding wild polar bears seems like a big step, like crossing some invisible line between the human world and the wild. But if we did it, it would actually join a long list of other actions we take to care for wild species. These situations are so common that the group of species we care for even has a name: "conservation reliant species." A recent book on the phenomenon defines the term like this: "A species is conservation reliant if it is vulnerable to threats that persist and requires continued management intervention to prevent a decline toward extinction or to maintain a population." These are species we cannot simply leave alone if we want them to persist. They are species that require intervention—at least for now. A 2010 analysis of the 1,136 species with recovery plans under the Endangered Species Act in the United States found that 84 percent require ongoing management.

Condors, even condors "in the wild," require constant, costly attention to avoid sliding into another decline. Other conservation reliant species are treated for disease. Others are protected by culling their predators. Some are fed.

The Kihansi spray toad lived in the delicate mist generated by Kihansi Falls in Tanzania. When a dam reduced the river's flow and the falls no longer created adequate mist to keep the toads moist, conservationists installed a sprinkler system.

The Oregon silverspot butterfly's seaside meadow habitat—no longer kept open by fire or elk—is now mowed and occasionally burnt, and native violets are planted every year for them to feed on. In addition, some butterflies are raised in captivity and released every year, a program that was partly funded by sales of Pelican Brewing's Silverspot IPA.

Grevy's zebras—"loiborkoram" in the Samburu language—are the largest wild animals in the horse family. There are fewer than 2,000 of them left in Kenya and Ethiopia. Overgrazing and drought have left their feeding grounds dry and barren, so conservationists put out thousands of bales of hay for them in the driest years. The southern white rhino is constantly protected from poachers and translocated from park to park periodically to avoid inbreeding. These are just a few examples out of many.

Once we get used to intervention, it is sometimes hard to know when to hold back. In some areas, grizzly bears are moving north at the same time polar bears are moving onshore as the climate warms, throwing these cousins into proximity. The two bears look quite different and lead very different lives, but they're actually closely related. Their lineages split less than 500,000 years ago. Now that they're bumping into one another, the odd hybrid bear is showing up. And because polars and grizzlies are so genetically similar, these hybrids are fertile. Here again, our cultural notions of discrete species and stable ecosystems are being challenged by the endlessly dynamic non-human world as it adapts to our influence.

What, if anything should be done about this? Should the encroaching *Ursus arctos* be shot, even though they are "climate refugees," because of the threat they pose to the purity of the species *Ursus maritimus*? Should the bears be left alone to mate how they please, to respect their sovereignty? Does it matter if the hybrids are less or more evolutionarily fit than the parents?

There's an amphibian case that could help us think this through. Endangered California tiger salamanders in Salinas Valley have been mating with barred tiger salamanders that were brought in as bait for bass fishing 60 years ago. The two salamander species had split off from a common

ancestor some 5 million years ago, but they were still able to mate successfully. Researchers initially saw the barred salamanders as a threat, and it was
presented as such when the California tiger salamander was listed as a
threatened species in 2004. But scientists found that the hybrids actually
have higher survival rates than either "pure" species. Whether you see this
as a tragedy of genetic pollution or a success story of genetic rescue depends,
the scientists wrote, on your perspective "on the merit of genetic purity as a
conservation goal."

What if the best way to save the polar bear is not to feed it but to let it
access the gene pool of its more flexible terrestrial cousin? Which would be
worth more to us, a pure white polar bear that we feed, or a grayish-
brownish northern bear that can live independently?

I think both allowing hybridization and feeding polar bears could be
ethically acceptable, if done right. A key criterion for the feeding program
would be the approval and participation of the Inuvialuit and, in other
populations, other Indigenous groups with traditional relationships with
the bear. I'd also feel more comfortable with it if the program would be
temporary, just to get the polar bear community "back on its own feet," as
Donaldson and Kymlicka say. But in keeping with my propensity to think
about conservation problems on millennial timescales, "temporary" could
be as long as 1,000 years.

Reading Palmer's paper, I kept imagining a day, in some future era long
after we are dead, in which climate change has not just been halted but has
been successfully reversed. The sea ice is back, and the seal hunting is good.
After generations of complete dependence on the feeding stations, scientists
have helped the polar bears in certain populations to relearn hunting skills,
a bit like the Adamsons did with Elsa. Polar bears have been relying less and
less on the supplemental food, and the decision is made to shut the program
down at long last. The humans feel good. They got these bears through the
hot bottleneck period. Now, out on the other side, their assistance is no
longer required.

A goodbye ceremony is planned. Project Polar Bear staff in thick parkas
sing a few songs to the last few bears, who they have names for and have

come to love as individuals. Local people come out with their kids; there's cake and punch. Staff bring out one last plate of food for the bears—by now some sort of nutritionally perfect plant-based chow. And then, they open the gates and gently shoo the bears out. As the bears wander away, casting long shadows on the ice, the humans wipe away their frozen tears, pack up, and go home.

10

The Arrow's Tip

To kill an animal is no small thing. To turn a thinking thing, a dreaming thing, a "subject-of-a-life" as Tom Regan says, into a still packet of meat and fur is a momentous act. It is also a common act. We've hunted for millennia. The hunting history of our species can be read in our cave art and in our bones, found mingled with those of the animals we killed, still scored with the marks of our butchering tools.

A few years ago, on a muggy November day, I went hunting for monkeys in Peru's Manú National Park—a huge swath of protected rainforest and one of the most biodiverse places in the world. We headed out around 6:15 AM, after a breakfast of fish and fish broth. Elias Machipango Shuverireni, who was carrying a palm-wood bow and arrows tipped with sharpened bamboo, was going to do the actual hunting. It would take all my energy and focus to keep up with him and not slip in the mud, trip over the buttressed roots of a fig, or walk face-first into a liana.

Manú National Park was created in 1973 and remains relatively unvisited by outsiders. It is not an easy place to get to—the most popular route involves a ten-hour drive down the Andes on a hair-raising road, followed by five hours in motorized canoe on the Alto Madre de Dios River to the mouth of the Manú River. From there, it is another two to five days' trip (depending on the water level) by motorized canoe to Machipango's settlement at Sarigemini.

Machipango belongs to an Indigenous group called the Matsigenka, of whom fewer than a thousand live in the park, mostly along the banks of the Manú River and its tributaries. They reside in a part of the park off-limits to tourists. All the park's Indigenous inhabitants have the right to harvest plants and animals for their own use, but they can't sell park resources without special permission, and they can't hunt with guns. Machipango and his family grow yuca, cotton, and other crops in a small clearing with several open, palm-thatched buildings on the Yomibato River. The family fish and gather fruit and medicinal plants. And they hunt, especially spider monkeys and woolly monkeys—favorite foods of the Matsigenka.

Things have been this way for a long time, but the Matsigenka are growing in number, which worries some biologists who love the park. What if their population continues to grow? What if they start using guns? Could the monkey populations survive? And without those species, which disperse the seeds of fruit trees as they snack through the jungle, how would the forest change?

Manú is traversed by tapirs, crowned by flights of scarlet macaws, veined with snakes. Ninety-two species of bats own the night sky. Butterflies are everywhere: cobalt and tomato-red scarlet knights, giant blue morphos, and tiny glasswings. And on every vertical and horizontal surface there are ants: leaf-cutter ants in lines like green sailboats, stinging bullet ants stalking the forest floor like lions. At night, the foliage beside the trails sparkles in your headlamp with what looks like pixie dust—the shining eyes of hundreds of thousands of insects.

Machipango has curly black hair and an intense gaze. On the day of our hunt he was wearing a green soccer jersey, shorts, and sandals made from old tires. As we crossed his fields and plunged into the jungle, we were accompanied by his son-in-law Martin, his daughter Thalia, and his teenage granddaughter Maria. Like Machipango, Martin was armed with a bow and arrows. Thalia wore a handwoven sling to carry back plants. Also with us was Glenn Shepard, an anthropologist who has spent more than 30 years working and living among the Matsigenka and may be the only person alive who can translate directly from Matsigenka into English. I was feeling very pleased that I convinced him to come along. We walked away from the river, across the family's field, and into the jungle.

John Terborgh, an ecologist at Duke who has studied in Manú for decades, for many years expressed the hope that the Matsigenka would leave the park—voluntarily, he emphasized—for the wildlife's sake and for their own economic opportunity. "Do I think there ought to be permanent settlements inside national parks?" he asked when I interviewed him at Cocha Cashu, his field station in Manú. "No. In this respect the U.S. model is a good one I am happy to endorse. Would you like to have farms and villages in Yellowstone or the Great Smoky Mountains?"

However, the Matsigenka's image of Manú includes the Matsigenka. Whereas Terborgh and other Western biologists come from a culture that separates humans from "nature"—both philosophically and as a conservation strategy—the Matsigenka see what many call "nature" as part of a humanized social order. They hunt monkeys; jaguars hunt humans. Key plants and animals have spirits and agency, just as people do, and there's no hard boundary between them. Many of their stories take place in a time when animals were all "people." Indeed, many animals still are people. Snakes are the arrows of invisible hunters, who hunt humans, which look like game animals to the invisible humans.

And as long as the Matsigenka don't use guns, Shepard says, their hunting isn't doing much harm. He and his colleagues asked 99 hunters to record the animals they killed, the ones that got away, and how long they traveled to find them. They found that the Matsigenka hunt five species heavily enough to theoretically reduce their populations over time—spider and woolly monkeys, white-lipped peccaries, and two birds, the razor-billed curassow and Spix's guan. But none of the species were yet declining. They also found that even if the Matsigenka population were to grow rapidly over the next 50 years, no more than 10 percent of the park would be depleted of spider monkeys—unless the hunters acquired shotguns. With guns, they could quickly empty the forest of monkeys within a day or two's walk of their villages. The Matsigenka understand this. They've seen or heard of the sad, empty forests outside of the park where people hunt with guns, and they know that a short-term meat boom would be followed by a prolonged absence of prey. This understanding may explain why they honor the gun ban despite very little enforcement. Shepard believes firearms would also spell the end of many of the rituals surrounding hunting, and perhaps this is another reason they have not taken up guns.

Five minutes in, we heard the calls of dusky titi monkeys, but the group didn't break stride. Titi monkeys are target practice for pre-teens. Another five minutes and we heard a troop of capuchin monkeys. Machipango paused for a moment, even raised his bow, but let them go. He was holding out for spider monkeys or woolly monkeys. We began a tour of fruiting trees, looking for evidence that our quarry had been feeding there, and we saw signs within another quarter of an hour. But the monkeys themselves were elusive, and we kept moving. Another hour or so, and Thalia said she heard spider monkeys—"osheto" in Matsigenka. Her face lit up with excitement as she described their position in a muted voice. The hunt was on.

Hunting monkeys with arrows is fiendishly difficult. The animals are in the crowded, complex treetops, up to 100 feet above the ground, leaping from branch to branch at high speed. The hunter has to shoot nearly straight up at an erratically moving target. Even hunting animals on the ground— boar-like peccaries, fat streamside tapirs—can be very tough in the jungle. As a result, the Matsigenka have an elaborate set of medical rituals to improve their hunting abilities.

To really prepare for a hunt, Shepard explains, it is good to take ayahuasca, a psychoactive combination of two jungle plants, made into a potent drink. The medicine makes the hunters vomit and purge themselves of harmful spiritual influences. It also provides a conduit of communication between the hunter and the spirits who control and own the game, which can be entreated for assistance. The next morning, they'll be purified and ready to go. Additional aid can be found by dropping the fluid from certain forest plants into the eyes, or by chewing special sedges called piri-piri. Still, success is never certain.

We followed Thalia's signals toward a troop of spider monkeys: dark, long-limbed shapes moving quickly far, far above us. Machipango bounded ahead. I struggled to keep up, feeling, as I so often did in Manú, clumsy and incompetent. It is hard to watch where you are going and keep an eye on animals so far above your head. Machipango caught up with a female spider monkey, took aim, and loosed an arrow. He missed and the monkeys bolted. If he'd had a shotgun, the monkey would have been dead.

Five hours into our hunt, Machipango and his family were still scanning the treetops, looking for monkeys. Sustaining this kind of intensive attention in such a visually complex environment is exhausting. Everything looked like a monkey—a dead bromeliad in the crotch of a tree, the shadow of a branch, a bird taking flight. My mind was so occupied with searching for monkeys, I didn't have the space to consider whether I really wanted us to find them. And I was physically tired. I figured we'd need to kill and eat a whole troop of monkeys to replace the calories we'd expended. But of course hunting fulfills cultural and social functions as well as secures calories. Along with specific nutrients, meat provides the hunter with social status, something valuable to give or trade, and—not least—the raw material for hunting stories.

Traveling along a ridge, we came across a mysterious, foul object—a wad of green leaves drenched in a dark liquid and covered with flies. Martin, Elias's son-in-law, explained that jaguars eat leaves and vomit them up, purging "just as we do, to be better hunters." Nearby, Machipango pointed out the wet stain of jaguar urine. Shepard translated in an excited whisper: "That piss is from *now*."

Machipango said that the jaguar had likely watched us as we passed that way half an hour earlier. As he spoke, the jungle exploded with deep, urgent cries: the jaguar-specific alarm call of an unseen troop of woolly monkeys just a few yards down the ridge.

The calls meant that there was a jaguar somewhere very close. The hunters froze. I felt an electric wash of adrenaline move from my feet to my scalp. With horror, I felt the terms of the day reverse. For hours, I had been searching the world for warm-bodied animals to kill and eat. Now another animal was very likely looking at me as a warm-bodied animal that he or she could potentially kill and eat. My flesh turned to meat around my tingling bones.

Machipango calmly sat down on a log and reached into a net bag made of tree bark. He pulled out a few roots of piri-piri and chewed them. Shepard has studied these sedges in some detail and found that the varieties they grow for medicinal use are filled with psychoactive ergot fungus. He tried one himself once, given to him when he had a bad headache, and found that it created a laser-like focus that lasted for hours. "Jungle Ritalin," he calls it.

Properly medicated, Machipango rose to his feet and, arrows in hand, plunged into the thick vegetation toward the cries of the woolly monkeys. He planned to take a woolly monkey—and the jaguar too, if possible. Better to kill it than merely scare it off, this close to his home. Jaguars kill children.

The rest of us waited, then crept down the trail as quietly as possible behind Machipango. A moment later, the heavens opened. Rain shot from the sky with the ferocity of a pressure washer. The noise of our movements now completely drowned out by the cacophony of a million glossy leaves being battered by raindrops, we sprinted off the exposed ridge and took shelter under thicker canopy to await Machipango's return. He appeared a few minutes later, smiling broadly, completely skunked by the sudden storm. Later, he would retell the story of the hunt, with lots of self-deprecating humor, over bowls of masato, tangy yucca beer.

The hunt was a failure. If Elias had infant children, a lack of hunting success would have meant the family's babies would be safe. After a hunt, the first priority of a Matsigenka hunter is to communicate to his wife which species he has killed. She will then prepare a medicinal bath with fragrant sedges and herbs specific to that species and use it to bathe any infants, who are at risk of being sickened or killed by the vengeful spirits of the prey. The Matsigenka live in a forest in which they are just one of many species trying to make a living—and not the only one with spiritual power.

Back at home, Machipango had no monkey meat to give his wife. But a baby spider monkey was warming itself by the fire. The Matsigenka love to tame forest animals as pets. When they do manage to kill a spider monkey, it often turns out to be a female slowed down by young offspring, and they sometimes bring the orphans home. Pet monkeys are often kept on a leash, fed and petted and fussed over. But once they grow to maturity, they are often let go, released back into the forest from which they came. If a member of the family later met and recognized a former pet when out hunting, it would be spared. But its children and grandchildren would presumably be fair game.

Thus the practice of pet-keeping seems to have the effect of restocking areas closer to villages with prey species, although when I asked about it, no one told me that this is why they did it. They gave me reasons that would be familiar to any cat or dog owner. They liked taking care of another creature. The pets kept them company. Many of the pets' primary caretakers were post-menopausal women, who were navigating an unfamiliar life without a bunch of human babies and kids to take care of.

This baby monkey was drenched to the skin, like the rest of us. I tried to give it some plantain, but it just looked at me skeptically. We joined it by the fire. The smoke rose above the papayas and floated across the river, out over the forest.

Terborgh and the Matsigenka have fundamentally different ideas about what an ideal Manú looks like. Where Terborgh revels in seeing monkeys "long accustomed to the benign presence of humans" up close at Cocha Cashu, Shepard tells a story about a trip far from the village with some Matsigenka friends: When monkeys approached them, unafraid, the Matsigenka men grew uneasy at the "unnatural" display. For them, humans are the predator of the monkey. It is normal that monkeys should fear humans.

———

Manú has a complex and often bloody history. In Incan times, people speaking Arawakan, Panoan, Harakmbut, and Tacana languages lived in the Manú region, taking advantage of the incredible difficulty—to the uninitiated—of moving through Amazonian rainforest, as well as their own skill as warriors to successfully resist Incan efforts to conquer them. The colonizing Spanish fared no better. Where people failed, though, viruses and bacteria succeeded. By the mid-seventeenth century, waves of diseases introduced by Europeans had swept the Amazon, killing untold numbers of people—and creating a "wilderness" that white people could admire.

At the end of the nineteenth century the Matsigenka, who speak an Arawakan language, all lived high up in the headwaters of the Madre de Dios and Urubamba rivers. Manú's riverfronts were occupied by other groups including the Yine (also known as the Piro—an originally derogatory name, which means "catfish") and the Toyeri. But then in the 1890s everything changed. The automobile had just been invented, and rubber for tires was selling at get-rich-quick prices. "Rubber barons" conscripted Amazonian native peoples to work at tapping the rubber trees in their forests. They hired guides from one tribe to help locate other tribes to raid for slave labor. They worked men to death and raped and killed the women. In the midst of this frenzy of violence and greed, rubber baron Carlos Fermín Fitzcarrald, 200 rubber tappers, and 1000 Indigenous workers pushed a riverboat over the isthmus separating the upper Mishagua River from the Manú river, opening up a huge new area to tapping. Once across, Fitzcarrald began trying to subdue and enslave the inhabitants of the Manú River—a thoroughfare then so densely populated that one local tribe called it Hakwei: The River of Houses. The battle saw casualties on both sides, but the Indigenous people got the worst of it. Hundreds died. The Manú was said to flow red from the carnage.

Brazilian explorer Euclides da Cunha, a witness to the Rubber Boom in Peru, described it like this: "Those strange white conquerors came in hurried rounds of slaughter of both men and trees, staying just long enough to utterly extinguish both, before seeking other paths where they would unleash the same chaos, passing like a destructive wave and leaving the wilderness yet more wild, more disordered."

The rubber boom collapsed in the teens of the next century. Fitzcarrald drowned in a boat accident. The Toyeri were almost entirely killed. The Yine, some of whom had worked for Fitzcarrald, eventually moved to towns like Boca Manú and Diamante on the Alto Madre de Dios River. So the

Matsigenka claimed the Manú riverfront, in a slow migration that began with the first missionary schools in the 1960s and continues today, as people drift down from the headwaters looking for axes, clean water, and education. Thus the political geography of Manú is neither primeval nor isolated. On the contrary, it was molded by the forces of the modern, globalized economy, in which technological innovation and consumer demand in one part of the world profoundly shape—and often damage—the lives of those at the source of the most valuable resources. Our "wildernesses" are just places where colonialism left the trees standing.

Today, at least several hundred Matsigenka live in Manú in small communities. Towns on the major rivers, like Tayakome, have schools, medical posts, and communal satellite phones. People here hunt, gather, and grow their own food, but they also play Peruvian pop hits on boom boxes, wear knockoff Crocs and T-shirts that say things like "Palm Beach," reserving their traditional garments for special occasions. Health in these villages was a problem for a while; the Matsigenka people used to move frequently and live in smaller groups, and exposure to introduced infectious disease was more limited. When they settled into permanent villages, people began getting sick. This recently improved, when the charity Rainforest Flow installed gravity-fed water treatment systems that run stream water through a filter of sand and pebbles and then, by pipe, to taps that deliver clean water to nearly every household.

Those who live away from the main rivers, in their steep headwaters, are much more isolated. They still wear handspun garments called kushmas and get by without money, metal, or personal names.

Shepard told me, "There are no demographic voids in the Amazon." That is, if all the people who live inside Manú were to leave—either by being forced out or by following educational and economic opportunities—someone else would come in: maybe illegal loggers or miners, certainly someone else who would be much harder on the forest than the current occupants, who like the jungle the way it is. And so with the Matsigenka as de facto guards, the park protects plants and animals that have been dramatically overharvested outside its borders, including woolly monkeys, turtle eggs, and cedro, a beautiful but valuable tree that seemed to disappear from the forest as soon as I stepped outside the boundary of the park.

Given the history of Manú, the idea of moving the Matsigenka or trying to restrict their right to hunt when they are doing so sustainably seems both cruel and ridiculous, just the latest way colonizers have decided to push Indigenous Amazonians around—punishing the very people who are protecting the monkeys from interlopers armed with guns.

A few days after almost becoming a jaguar's dinner, I interviewed Mauro Metaki, a genial mission-educated schoolteacher in the village of Tayakome. We sat on the open first floor of his house, smelling the fragrant grass around it. He told me how he loves to watch the oropendola birds that visit his wild palm and the squirrel monkeys that come to eat his bananas. Across the river, he hears the soulful hoots of the howler monkeys. He was happy to be interviewed by me, he added. He wanted the white people to know that the Matsigenka live here. "There's a park," he said, "but there are also people living here—right in the middle of it."

Metaki has been urging the Matsigenka communities on the river to reject new road building and job offers in mining, logging, and agriculture outside the park, where they are often exploited. One Matsigenka man from Yomibato told me he gathered Brazil nuts for "a white man named John" and was paid the equivalent of $115 dollars for three months of work.

Instead, Metaki wants trade schools established here, inside the park, so the people can remain close together and close to the river. "We live here," he said. "Sure, we hunt and fish, but we take just a little to feed our families. We know how to take care of the forest."

In the Matsigenka's world, killing animals isn't an ethical conundrum, it is a mandatory part of staying alive. They kill to eat and also to protect their children. And yet they don't see themselves as masters or overlords of other animals. They see themselves as just people in a jungle filled with people of different species—and their willingness to both eat monkeys and invite them to sit around the fire speaks to the complex fluidity of their relationships.

———

Indigenous people are not all the same; the idea that there's one Indigenous hunting culture is obviously absurd. But there's one theme that keeps coming up again and again—the notion of hunting as a reciprocal practice. Anthropologist Paul Nadasdy, who studies First Nations peoples in the Arctic and Subarctic describes a "reciprocal exchange between hunters and other-than-human persons." "In their view, fish and animals are other-than-human persons who give themselves to hunters in exchange for the hunters' performance of certain ritual practices. These practices vary across the North—as well as by animal—but they commonly include the observance of food taboos, ritual feasts, and prescribed methods for disposing of animal remains, as well as injunctions against overhunting, and talking badly about, or playing with animals." Hunting in such societies should not be viewed as a violent process whereby hunters take the lives of animals by force," he continues, "but, rather, as a long-term social relationship between animal-people and the humans who hunt them."

"Hunting was not only a display of human prowess but also an opportunity to acknowledge the reciprocal relations linking men and animals," writes Virginia DeJohn Anderson of Powhatan hunters in what is now Virginia in the early 1600s. After a successful outing, hunters would offer thanks to spirits who had allowed them to be successful. A kill did not just rely on their skill or stamina, but also on successful "negotiation with prey animals' spiritual protectors." These relationships were conceived of as "mutual support," not domination or exploitation.

Some Mayan hunters in the Yucatan Peninsula in southeast Mexico ask formal permission to take game animals from the Lords of the Animals through a ritual called Loojil Ts'oon. The hunters prepare a special soup made of both game and rooster meat for the god Sip. As the officiant at one such ritual explained to researchers, "The rooster is the offering for the Lords of Animals. It is an exchange. You trade the rooster for the wild animals." The next day, the hunter takes the jaws of 13 killed deer or peccary and returns them to the woods so that they can be reborn.

If hunters fail to complete this ritual every 13 kills, they may receive warnings that they are "in the red" with the spirits. They may run into venomous snakes, trip and fall, have rocks hurled at them by little forest spirits called aluxo'ob, or get sick. Eventually, Sip may appear to them in the form of a deer to warn them that they are courting disaster.

Robin Wall Kimmerer speaks of "honorable harvest" of both plants and animal persons in contemporary Indigenous thought in North America. "Killing a *who* demands something different than killing an *it*," she writes. "When you regard those non-human persons as kinfolk, another set of harvesting regulations extends beyond bag limits and legal seasons." The main laws of honorable harvest are to take only what you need, to ask permission, to accept no as an answer, and to take life with gratitude.

How do you ask a deer or a fish permission to kill and eat it? Robin Wall Kimmerer quotes one hunter who only shoots at a deer if the shot is absolutely perfect. By standing still out in the open with its flank exposed, the deer gives him permission. "I know he's the one and so does he," the hunter says. "There's a kind of nod exchanged." Whether or not you accept the more spiritual aspects of such a cross-species relationship, only taking animals that seem to offer themselves does reduce the chance that animals will be injured but not killed outright, which might create a prolonged period of pain in the fleeing animal.

Of note in these accounts of Indigenous hunting is that the hunters arguably treat their quarry with more respect than the average "animal-lover" might. They see the hares, monkeys, and deer not as lovable, cute innocent forest creatures that must be protected at all costs, but as people, as equals. Their relations with them are not paternalistic—but neither are they antagonistic. Their relations are reciprocal. For the life taken, something is given, either through a prescribed ritual, through hours of intense effort, or simply as deeply felt gratitude.

———

Most non-Indigenous people in the Western world buy meat from a grocery store, butcher, or farmer. But some who have the means to get their meat in a Styrofoam tray still choose to go out armed into the woods and fields. Why? Many who oppose the industrial food system as cruel to animals believe that hunting is a more ethical way to get meat. And in ecosystems where top predators like wolves have been lost, human hunting can step in to regulate the numbers of prey so that they won't overgraze their home and then die of starvation. In some suburbs, the deer population is so dense that teams of sharpshooters are called in to cull numbers. But then the meat is often "wasted" by being thrown away (although some creatures will eat it eventually, even if they are mostly insects and bacteria).

I know many "eco-hunters" who kill to support the management of populations of deer and feral pigs as well as to feed their families. The founder of conservation biology, Michael Soulé, told me he would hunt one elk a year. When the season was almost over and he hadn't gotten his elk, he'd get anxious. My own father-in-law drives a Prius with the vanity plate ECO-HUNTR. My first real date with my husband was in 2003; he cooked me venison braised with dried cherries. It was delicious. After I polished it off, he told me his father had found the deer dead by the side of the road, hopped out of his car, and quickly and expertly removed the backstrap. When I didn't look revolted, I suppose I passed some kind of test. I've eaten a fair amount of roadkill in the years since.

My first time hunting was in college, in 1999 or so, when I went deer hunting with a big family of liberal Texans. My attitudes toward the practice at the time were pretty typical of urban, lefty environmentalists: I thought it was a barbaric pastime for dudes who needed to prove how macho they were. But by actually going hunting, I learned a few interesting things. Firstly, certain kinds of hunting require waiting silently, without moving, for long periods. In all my years of hiking and camping, I never spent so much time simply sitting in the woods and observing as I did sitting in a tree stand in Texas with a loaded rifle across my lap. After several hours, a certain zen-like trance seemed to descend, and I felt the focus of looking for deer completely blot out all other thoughts.

I didn't shoot a deer that day, but some of the others in my party did. I have never killed an animal while hunting, but I have intended to, and I have shot *at* some ducks, which were felled by my more experienced brothers- and sisters-in-law.

Hunting is interesting to children. When my nephew shot a rabbit, all his little cousins crowded around the animal's still body. They all wanted to touch it, to look at it, to hear the story over and over from their older cousin. The mood was solemn but focused. There seemed to be something evolutionarily ancient in those responses.

The moment reminded me of my mother's salal bushes at her house in Seattle. When my daughter was two, and my son an infant, my mom showed my daughter that the salal berries were edible. The little orbs are as black and shiny as patent leather. They taste bland but pleasant. I was astonished when, a full calendar year later, we returned to my mother's house, and my daughter taught her toddler baby brother to eat them. Some primeval brain circuitry had held on to the information about salal berries with a tenacity carved by millennia of hunger. And some animal instinct prompted her to pass this knowledge to her brother. Foraging and hunting feel like old ways of being human.

I've eaten ducks, pheasant, caribou, elk, moose, chukars, turkeys, trout, and salmon that were personally killed by someone I knew. These animals lived largely autonomous lives up until the moment of their swift deaths. Wasn't that better than eating an animal raised for meat and kept confined? And instead of outsourcing their deaths to underpaid, exploited slaughterhouse employees, my friends and family did it themselves, which seems arguably more responsible.

Getting your meat from outside the industrial food system is also better for the environment. Wild game animals aren't fed on tons of grain that requires excessive water, land, and fossil-fuel-based synthetic fertilizer. They aren't clustered in "concentrated animal feeding operations" that produce toxic and terrible-smelling lagoons of manure.

Of course, the counterargument is that one does not need to eat any meat, factory farmed or felled in the forest. That is true. I have the utmost respect for vegans. Making that decision for oneself can be one way to live your animal and environmental ethics every day. And I'm thrilled at the expansion of plant-based options to take the place of hamburgers, cheese, milk, and other staples of the Western diet. For many reasons, including animal welfare and the environmental footprint of beef, pork, and chicken, eating less meat or none at all can be a deeply ethical act.

Even vegan food does have effects on non-human animals, however, insofar as agriculture takes up space that was once habitat for wild animals, and animals such as insects and rodents are controlled on the farm and along the supply chain. And of course eating plants does mean killing living organisms that—though they may not have a consciousness like ours—certainly sense and respond to their environment and have goals. There is no way to fully opt out of ecological existence.

There's also a difference between choosing to go vegan oneself and demanding everyone do so. There are plenty of people—especially Indigenous people—who hunt and eat meat as a core expression of their cultural identity.

Philosopher Val Plumwood's sense of the importance of these cultural hunts and our role in the food chain led her to reject arguments for universal veganism as "aggressively ethnocentric, dismissing alternative and indigenous food practices and wisdom and demanding universal adherence to a western urban model of vegan practice in which human predation figures basically as a new version of original sin." She argues that refusing to hunt or eat animals can actually alienate us from the rest of Earth's species.

Plumwood's approach to ethically inhabiting our ecological role was partially inspired by being attacked by a saltwater crocodile while canoeing alone in Kakadu National Park in Australia in 1985. The crocodile swam toward her and rammed her canoe, over and over. "As I paddled furiously, the blows continued," she writes in a retelling of the incident. "The unheard of was happening; the canoe was under attack! For the first time, it came to me fully that I was prey." Plumwood decided to try to leap into a tree to escape the attack. "At the same instant, the crocodile rushed up alongside the canoe, and its beautiful, flecked golden eyes looked straight into mine." She jumped; the crocodile lunged after her. "I had a blurred, incredulous vision of great toothed jaws bursting from the water. Then I was seized between the legs in a red-hot pincer grip and whirled into the suffocating wet darkness."

The crocodile commenced a "death roll," spinning sideways in a technique intended to wrench prey so forcefully that it usually won't just be stunned—it'll be dismembered. Plumwood described it as "essentially, an experience beyond words of total terror . . . a centrifuge of boiling blackness that lasted for an eternity, beyond endurance." The croc rolled her twice. She broke free; it seized her again in its massive jaws and rolled her again. She broke free again, and this time managed to scramble up the bank.

Grievously injured, she managed to walk nearly two miles, eventually crawling to the edge of the marsh, where she was rescued. In 2008, she wrote that ever since that day, "it has seemed to me that our worldview denies the most basic feature of animal existence on planet Earth—that we are food and that through death we nourish others . . . Dominant concepts of human identity position humans outside and above the food chain, not as part of the feast in a chain of reciprocity."

How can predation be "reciprocal" when one party ends up dead? Hunting rituals like returning the jaws of prey to the forest seem to even up the score in a spiritual sense, but what if you do not believe in the Lords of the Animals? Plumwood says you cannot look for a reciprocity between both parties in that moment of death. But insofar as both have participated in the "systems of flow and exchange that nurture all life," there's a way in which both parties have benefited in a broader sense. When the mouse is killed by the hawk, it suffers and dies, but before that it was nourished by fatty seeds growing on plants, plants that built their bodies with carbon breathed out by the hawk, nutrients added to the soil through hawk excrement, eggshells, and the bodies of dead hawks. Everyone benefits from these cycles.

Death is frightening. It is unsurprising that we recoil in horror at the idea that humans are also food for other species. But everybody gets eaten, sooner or later: plants, deer, wolves, humans. We all end up as food in the end.

Those who dislike hunting often see it as domination and cannot imagine a way to kill and eat another animal without treating that animal as a means to an end. But as we've seen in Manú, there's another way. You can treat your quarry as a powerful being that you must negotiate with. You can have a reciprocal rather than dominating relationship.

Plumwood, though, also looks askance at white, privileged people engaging in hunting as mere recreation. How would she judge me, visiting family on Whidbey Island, Washington, sitting down to eat a portion of grilled salmon my sister-in-law and her family caught? I can't ask her, alas, as she died in 2008, becoming, as she would say, part of the feast. But I think she would want to know whether my family killed the fish with respect and gratitude, whether they made sure not to waste any of it. She might want to know my thoughts on reintroducing grizzly bears to Washington State, since embracing ecological embodiment means being willing to also share the world with those that would see us as prey.

Although I have never had an encounter with a predator as intense as Plumwood's, I did, on that Amazonian ridge top, have a brief but crystalline moment of understanding that my body could be food. As the woolly monkeys around us hollered their jaguar alarm cries, I remember feeling grateful to them for "warning us." Given that we were out looking to hunt them, the notion that we were on the "same side" seems absurd, but for that brief moment, my position in the forest had changed from potential predator to potential prey. I was no longer Emma Marris, environmental writer, defined by my mental activity and the words I've strung together over the years. I was a body that could be food, a packet of energy in a jungle full of hungry eyes.

11

Bloodshed for Biodiversity

Karl Campbell is a middle-aged, medium-sized Australian with a five-day beard and an intense gaze. He seems perpetually coiled, even angry, when at rest. He's smiling and relaxed only when his body is in motion—preferably fixing something, building something, or killing something.

Campbell lives in the Galápagos Islands, full time. Far off the coast of Ecuador, the island chain is a place of almost pure geology and biology, with no local gods, no legends, no human history before 1535, and no permanent residents before 1805. Human changes to the land are still fresh enough that much can be undone with work and money—and killing. Most of the original species remain, from marine iguanas shooting salt snot from their nostrils to waved albatrosses gliding on eight-foot-wide wings, scanning the sea with wet eyes like black tapioca balls.

This is where Charles Darwin collected the first inklings of what would become his theory of evolution. The slight differences in species from island to island suggested to him that perhaps they were all descended from a common ancestor. Tourists flock here now to see these species, to pay homage to Darwin's great discovery, to experience "nature" in a rawer form.

I flew into the airport on Baltra Island, and made my way to the archipelago's biggest city: bustling, knicknack-riddled Puerto Ayora, on Santa

Cruz Island. The next day, I took a ferry to Floreana, a volcanic island of 173 square kilometers with a human population of just 100—most of whom both farm and work in the tourist trade. Campbell, who works for an organization called Island Conservation, picked me up at the dock, upon which sea lions lolled and ruby-red marine iguanas regarded me with disinterest. Campbell wanted to show me some tortoises that live up the road in a sanctuary. We hopped in the back of a truck, which functions as a kind of tourist shuttle, and headed up the side of the island's central volcano.

Campbell is extremely driven. He prizes efficiency and logic. He has done the math on stopping extinctions. The roughly 465,000 islands in the world represent just 5.3 percent of the Earth's landmass, but 75 percent of bird, mammal, amphibian, and reptile extinctions since 1500 have been island species. Why? Animals on islands evolve in unique directions, especially when there are few or no predators to worry about. Birds may lose the ability to fly and the instinct to flee. Island animals don't need to spend so much energy on fighting for survival. The rules are different on islands.

In recent centuries, though, humans have changed the rules. Firstly, we hunted out many island species ourselves. It is all too easy to go overboard when your prey just sits there blinking at you when you try to hunt them. In this way, we lost the great auk, a majestic black-and-white flightless bird that lived on rocky islands in the Atlantic, and the famous dodo, which lived only on the island of Mauritius. In this way, we lost the Falkland Islands wolf, the Caribbean monk seal, and nine species of New Zealand moas, those long-necked flightless birds whose bones I saw in a natural history museum. This is how the Galápagos lost the Floreana tortoise, the last specimen of which was turned into soup in about 1850.

Secondly, we made islands functionally less remote by installing seaports and airports and visiting them and leaving and coming back. Just as we have reconnected continents separated by plate tectonics with our globalized trade and travel networks, reuniting Pangea in practice, so have we pulled archipelagos together and closer to continents. In the twenty-first century, almost nowhere is really far away. And where we go, so go our pets, our crops, our livestock, our medicines, our synanthropes and our kleptoparasites—the whole human entourage. Island animals often just as defenseless against smart, flexible killers like cats, foxes, rats, and snakes as they are against us.

The placid reptiles at the Asilo de la Paz—the "Peace Haven" sanctuary—are not strictly speaking native to this place. They are retired pets brought from other islands along with their offspring—a proxy for what is missing. The Galápagos tortoises from Floreana had "saddleback" shells with a high arch in the front, which allowed them to stretch their necks way up high to nibble on cactus fruits several feet above the ground. The tortoises I am looking at now mostly have the classic dome shape, because their ancestors came from islands with more low-lying vegetation. Campbell and I watch them eating iceberg lettuce as if they have all the time in the world—and with a life span of well over 100 years, why not be leisurely? Their bumpy limbs and seemingly sour expressions are inexpressibly charming, but they are penned up, not free-roaming.

There is some hope of "rewilding" the Floreana tortoise. That may sound bizarre, since I just said it was extinct. And it is. But in 1994, a biological expedition on the remote Wolf Volcano on Isabela Island, more than 100 miles from where I stand, found saddlebacked tortoises. Later, scientists determined that these individuals were hybrids—a mix of different island species with some Floreana tortoise parentage. Apparently, this population was descended from a living meat cache set up by whalers or buccaneers in the 19th century. The seafarers nabbed tortoises from across the islands and stashed them on Wolf as emergency rations. In 2015, 32 tortoises with pronounced saddles were captured and airlifted back to Puerto Ayora. The plan is to breed them carefully, maximizing the genes from Floreana, and eventually return them to Floreana.

Before they can be set free to live out their slow lives in the Equatorial sun, though, Floreana needs some ecological adjustments. Rats eat Galápagos tortoise eggs and babies. And Floreana is crawling with rats. Campbell would like to change that.

———

Because the changes to the Galápagos are so fresh, conservationists see a real opportunity to prevent extinctions there through vigorous action, to do the right thing by at least one ecosystem. At the same time, the archipelago exhibits the characteristic conservation problems of islands, including introduced predators and vulnerable native species that sometimes seem determined to perish. And Campbell's approach to fixing it is the standard approach for islands: kill the interlopers, undo new ecological dynamics, and try to prevent extinctions at all costs. Island conservation is all about killing these days. What I wanted to know was whether preventing these extinctions was always worth the price in blood.

When humans first came to the Galápagos, they brought beasts of burden, animals for meat, and the clever and voracious rat, hidden in the holds of their ships. The animals of the Galápagos, like island species everywhere, had let down their defenses over evolutionary time and simply could not cope with these bulldozing newcomers. Even when the animals humans brought didn't eat the native fauna, they did damage in other ways. Free-roaming goats ate so many plants that one estimate claimed that 60 percent of the Galápagos' 194 endemic plants were threatened with extinction—not to mention the islands' giant tortoises, which were starving to death with no plants to eat.

Rats have already killed off all the populations of the ironically named Indefatigable Galápagos mouse on Santa Cruz Island. In 2005, a single cat was found to be responsible for eating seven endangered Galápagos penguins every month at one breeding site on Isabela Island—a rate of decline the colony could not have sustained if researchers hadn't killed the cat. Rats, cats, and dogs exiled the Floreana mockingbird—a chocolate brown bird with a perky tail—to two minuscule offshore islets.

This pattern is not unique to the Galápagos. Non-native species are implicated in 62 percent of amphibian, reptile, bird, and mammal extinctions (although many had more than one cause listed). But importantly, of those cases where "alien species" were listed as a driver of extinction, a whopping 86 percent of the species lost were "island endemics"—occurring only on islands.

It is important to note that introduced species are much less likely to cause extinctions on continents, because there's time and space for the native species to adapt to the new presence. Newcomers may well cause declines in *abundance*—the sizes of native species' populations. For example, free-ranging domestic cats kill up to 4 billion birds and 22.3 billion mammals every year in the United States, according to one analysis. But as of 2020, cats haven't caused any extinctions in continental North America. Meanwhile, on islands—even islands as large as Australia—cats have driven dozens of species extinct. Cats have been a factor in 63 extinctions—every single one of which was Australian or an "island endemic." In fact, *all* the species driven extinct since 1500 by non-native animal predators were either Australian or island endemics.

To me, this suggests that our thinking around "invasive species" needs to be fine-tuned. Instead of a paradigm where we see all "foreign" species as malevolent invaders that should be considered threats to ecological integrity unless proven otherwise, maybe we should instead see islands species as particularly vulnerable to newly arriving species.

Indeed, the overall concept of the "native" has some fundamental problems. It derives from precisely that frozen-in-time idea of "ecosystem integrity" that, as we've seen, is riddled with conceptual shortcomings. Ecologists have spent decades assigning "native ranges" to species, usually based on where they were when the first white scientist showed up to take notes. These ranges are pegged to an arbitrary point in time, a moment in the long evolutionary and geographical journey of a particular lineage.

For example, ancestors of the Virginia opossum evolved in South America, then entered North America after the continents joined up, about 800,000 years ago. They've been slowly ambling north ever since. By the 1600s, they had made it to Ohio and by the 1920s, opossums had made it to southern Michigan. Today, you can find them as far as southeastern Ontario. This recent range expansion has been enhanced by climate change and human changes to the landscape, but they are doing it without being physically moved by people. So what part of the Americas should count as their "native" range?

In the Northern hemisphere, many species have shifted ranges over and over as glaciers have advanced and receded over the millennia. Every single species in Canada today arrived there less than 20,000 years ago, because before that the entire country was covered by a solid block of glacial ice. The same goes for much of Northern Europe. As the climate began to warm, species began to move north—though some moved east, west, or even south—and they all moved at different speeds. Arctic ptarmigan used to roam Central Europe; Greenland collared lemmings once lived in what's now the United States.

When humans move species, those new areas never count as part of the "native" range, because of the fallacious idea that humans aren't part of nature. This makes things confusing when those movements happened so long ago that it has become impossible to untangle human influence. Kukui, or candlenut, a tree with a nice fatty seed you can eat, burn for fuel or light, or turn into moisturizer, was so useful to people that they moved it all over the Southeast Asian tropics and Oceania. No one knows where its "native" range was. It is now the state tree of Hawai'i.

The camel family evolved in North America, then spread out. The ancestor of *Camelus dromedarius* ended up in the Middle East and North Africa, where it was domesticated about 2,000 years ago. Then the wild dromedaries went extinct. Today, descendants of domestic dromedaries brought to Australia by colonists as work animals in the 19th century roam free in the Outback. They are considered "invasive." Government sharp-shooters regularly cull them from helicopters.

Climate change is shifting species all over the planet—although they still rarely cross oceans without human aid. But as continental flora and fauna shift polewards, the idea that everything "should" stay in its native range becomes increasingly untenable. In North America, beavers are moving into the Arctic tundra, completely reengineering the hydrology of the landscape.

In April 2020, an ecologist named Mark C. Urban published a paper in the journal *Nature Climate Change* entitled "Climate-Tracking Species Are Not Invasive," seeking to differentiate species that are moving on their own in response to rising temperatures from those moved by people. "The same climate-tracking species arriving and disrupting a local community might also be threatened in their original range. Preventing shifts in species and ecosystems in favour of local, historic patterns is not only likely to be futile, but could cause range collapses or extinctions at broader scales . . . During climate change, we should keep nature alive, even if it happens to be in a different place."

I think Urban could have gone even further. Plenty of species that humans moved, such as the dromedary, are endangered or extinct in their "native" range. Don't they deserve to be kept alive too?

The idea of ecosystems as "stable" and of humans as forces that can only destroy "naturalness" come together in the war on "invasive" species. I honestly hate the term. It suggests the incoming organisms are showing up intentionally, with actively malicious intent. It bears repeating: So-called "invasive" species do not know they are in the "wrong" place. They are not trying to cause harm. They are just trying to live. If their success in a new location causes undesirable effects, there may be a good argument to try to move them, maybe even kill them. But they are not morally blameworthy. When an animal is trapped and killed simply because it is not native to a place, and despite the fact that its presence isn't causing any real problems, our instinct to protect "ecosystem integrity" or to keep the place "natural" is leading us to some morally dubious decisions.

———

So the places where we are most likely to see compelling arguments for killing non-native species are islands, because of the special vulnerability of the species that inhabit them and their restricted size—an introduced predator can kill them all before they have time to adapt. These island eradications are Campbell's specialty. It's a grueling job, preventing the catastrophe of irreversible extinction with a tide of blood. He kills goats and rats and other human-introduced animals that threaten rare island creatures, but his tools—traps, long-range rifles, and poisons—are brutal, deployable only on a small scale, and often all too indiscriminate. To excise the rat, say, from an ecosystem, conservationists typically distribute poison that can kill many species.

Around the world, conservationists routinely and increasingly kill free-ranging animals to protect endangered species. One particularly massive operation was undertaken to save rare birds on South Georgia Island in the South Atlantic. There, conservationists dropped 300 metric tons of poison bait on the sub-Antarctic island to exterminate rats. The operation cost $13 million and was declared a success in 2018. The birds are already bouncing back, with the endemic South Georgia pipit—a speckled songbird—exploding in numbers after the rats were dead.

Some conservationists aren't just willing to kill animals to save species. They are willing to use other animals as the weapon. Consider the "death row dingoes" of Pelorus Island. In 2016, several dingoes were set loose on this Australian island to kill introduced goats that were allegedly eating the native plants to nubbins. The biologists who planned the project didn't want dingoes on the island either, however, so they planned to shoot the canines once they had killed all the goats. But dingoes can be tricky to hunt down, so just in case, the biologists borrowed a plot point from John Carpenter's 1981 film *Escape from New York*: The dingoes were implanted with capsules of a poison called 1080 that would break down over time, eventually killing them.

Conservation biology has, from its beginnings, always explicitly stated that its concern is with populations, not individuals. In his famous 1985 paper where he lays out the core values of conservation biology, Michael Soulé rejects the inclusion of any "normative postulate" concerning individuals. "It may seem logical to extend the aversion of anthropogenic extinction of populations to the suffering and untimely deaths of individuals because populations are composed of individuals," he writes. "I do not believe this step is necessary or desirable for conservation biology. Although disease and suffering in animals are unpleasant and, perhaps, regrettable, biologists recognize that conservation is engaged in the protection of the integrity and continuity of natural processes, not the welfare of individuals."

There's no comprehensive figure for how many animals conservationists kill each year, but it is almost certainly in the hundreds of thousands at least. Australia culled approximately 211,560 cats in twelve months in 2015–2016 alone, a pretty typical annual death toll. Also in 2015, conservationists eradicated red deer from the Fiordland Islands of New Zealand, rats and mice from South Georgia near Antarctica, rabbits and mice from the Madeira Islands of Portugal, a songbird called the Madagascar red fody and the common myna from islands in the Seychelles, the Polynesian rat from the Tongatapu Group of Tonga, mule deer from the Channel Islands of California, sheep from the Kerguelen Islands in the Indian Ocean, rats and cats from the Tuamotus and Gambiers in French Polynesia, and rats from the Kelp Islands of the Falkland Islands. Since 2015, the number and scope of these eradications has only increased. These days, I'd guess it is in the millions every year.

Arguing whether killing sentient animals is bad may sound absurd, but there's a real philosophical disagreement between those who see instant, painless killing as morally neutral for some animals and those who do not. As we saw, Peter Singer falls into the former camp. Singer is focused on sentient creatures satisfying their preferences. Some, he says, don't really understand that they are an individual with a past and a future; they don't make plans for the day after tomorrow. Therefore, he says, "a being which cannot see itself as an entity with a future cannot have a preference about its own future existence." If you kill them quickly and neatly, there's no wrong done.

Others argue that even if some animals don't prefer to exist in the future, killing an introduced rat or cat deprives them of years of well-being they might have had if you had not killed them. This focus on lost well-being brings up another way all deaths might not be equal. The premature deaths of elephants in zoos strikes us as sad and perhaps unjust in part because these long-lived animals are deprived of many years of life. Thus the deaths of long-lived animals or young animals with their "whole lives ahead of them" might be worse than the deaths of short-lived animals or animals that are nearing the end of their days anyway. The mice and rats that are so often the targets of conservation killing do have short lives compared to the giant tortoises and large seabirds they are eating. Rats live a year or two, maybe seven at the very most; albatrosses can make it to at least 70; giant tortoises routinely see their 100th birthday and some have approached 200.

Pain and suffering are arguably much more straightforward than death, ethically speaking. On the whole, they are bad for any sentient creature. Some conservation killing—via well-placed traps or at the hands of a very good shot—is nearly instant and painless. A lot of it, though, is not.

On Floreana, the rats' destruction will be brought about by a carpet-bombing of poisoned cereal pellets: Some 300 tons will be dumped from helicopters, enough to kill every rat on the island. The poison Campbell will use is called brodifacoum. It is an "anticoagulant poison," which means that it stops the rats' blood from coagulating. The rats bleed from internal organs and sometimes their eyes, nose, gums, and other orifices in the course of about six agony-filled days. In lab tests, rats that ate the poison hunched up, stopped moving, and bled externally. Some became completely paralyzed. It took them about a week to die. During the last 11 hours or so before death, they "remained conscious but unmoving . . . except for some occasionally pushing or pulling themselves along the floor." Unsurprisingly, this type of poison is consistently rated as causing the absolute most suffering out of all pest-control methods, but it works. Part of why it works is that the rat doesn't feel bad right after eating the bait. It takes hours or even days to start bleeding out. So the clever mammals never learn to associate the tasty cereal pellets with sickness, as they often do with faster-acting poison. The slow death is a feature, not a bug.

———

When Campbell was a kid there was a show on Australian TV called *Wombat*, named after a native Australian animal that looks a bit like a furry ottoman with a teddy bear face. Once, the show featured a segment on the captive breeding of rare birds. At age seven, Campbell was entranced. He began to practice breeding birds at home—quails, doves, parrots, finches. He liked that it was an active way to help endangered species.

Although his parents "probably had the resources" to fund his education—Campbell says he never asked—he didn't want to rely on anyone, so to pay for college, Campbell joined the Australian Army Reserves, hopping on a bus to go to recruit training just 12 days after his high school graduation. There he learned to shoot and repair vehicles. After a year of full-time soldiering, he started college in Wildlife Management. To support himself, he worked construction. During college he took time off to recover from an elbow injury he says he got from a combination of rock climbing, rugby, and changing truck tires. After that he spent a month in Malawi arresting poachers, "to clear my head."

Linda Cayot, former project coordinator for Project Isabela, a goat eradication drive on three Galápagos islands, recalled that when she picked Campbell for an internship with the organization back in the late 1990s, one of his virtues was a "certain macho army roughness." Campbell had learned to shoot firearms and repair vehicles in the Reserve. He was well suited to the demands of the work on the islands: Once he slashed open his thumb and had a friend stitch it up in the field; another time he came back from a visit to Wolf Volcano with most of the skin on his feet peeling off. He didn't bother to mention it.

For Project Isabela, Campbell shot goats with semiautomatic .223-caliber AR-15 rifles, mostly from helicopters, occasionally on foot with dogs. But he quickly recognized the imperfections of these methods. You could never get those last few goats, and they would quickly breed and restock any island. He came up with a strategy for inducing sexual receptivity in females in order to lure other goats out of hiding so they could be shot. The resulting "Mata Hari" goats were a big success and propelled Campbell to a kind of fame in the conservation world—and earned him a PhD from the University of Queensland.

In 2006 Campbell went to work for Island Conservation, taking his skills beyond the Galápagos. He has helped rid San Nicolas Island, California, of feral cats; Choros Island, Chile, of rabbits; and Desecheo Island, Puerto Rico, of rhesus macaques. Campbell had found his true calling. He told me he realized that breeding endangered species isn't much different from collecting antiquities unless there's some real chance of reintroducing them to the wild someday—unless the major threats are dealt with. "I was doing all this work breeding birds and I should have been learning to kill shit," he says.

Campbell is now focusing on Floreana's cats and rats, aiming to eradicate them completely by 2021. Once the cats and rats are gone, the Floreana mockingbird could be brought back from those two offshore islets to the place for which it was named. Removing cats and rats will also clear the way for the return of saddleback tortoises with at least a healthy chunk of Floreana tortoise DNA. It will safeguard other threatened species as well: the Galápagos petrel, 60 percent of which nest on Floreana, and twenty species of endemic land snail.

The biggest problem for Campbell is that brodifacoum poison also kills farm animals and some native animals. So he can't just dump it on the island willy-nilly. Everything that people want to keep alive must be kept indoors, moved off island, or otherwise protected from the rain of death. And so Campbell has to work with every single household and farm on the island to prepare.

The morning after our tortoise visit, Campbell and I hopped in a local farmer's battered Toyota Land Cruiser and headed for the highlands of Floreana. Rats are no friends to farmers either, and Campbell pointed to some corn that had been nibbled away by sharp rodent teeth. Campbell estimates the famers here lose as much as 40 percent of their crop to rodents. Farmer Claudio Cruz, a smiling middle-aged man with reading glasses pushed up on the top of his head, showed me the poison bait he strings like a necklace of pearls around his crops. Even with all this poison, rats burrow into his yucca, he says, hollowing them out from the inside. "A Floreana without rats would be marvelous," Cruz said in Spanish.

On the day of my visit, Cruz was showing off his spread to Campbell and two farmers from another Galápagos island, San Cristóbal. Both were named José; they were shopping for calves. After showing us an ingenious compost system of his own design, Cruz called in his cattle. After a few minutes of hollering, the cattle emerged from banana groves, brown and white, single file. They were attended by white cattle egrets, a species that has managed to conquer the world on its own, spreading along with cattle from continent to continent, a quiet avian version of us. As the cattle lumbered through the grass, a panicked rat made a run for it, dashing under Cruz's truck.

After looking over the calves the group stopped by an orange tree heavy with fruit and one of the Josés climbed up and began tossing fruit down to Campbell, who packed them up in a feed sack. Nearby, Cruz had parked a couple bright red shipping containers up on blocks—one a gift from Island Conservation, one he bought himself. They will protect his animal feed when the poison comes.

Island Conservation will also build coops, sties, and stables for the island's chickens, pigs, and horses. It will buy "sentinel pigs" that will live outside the sties and be slaughtered at intervals so their livers can be tested for poison. The other pigs won't be able to emerge until the sentinel pigs' livers are clear. This might take three years. Parents will have to keep close watch over small children lest they eat pellets off the ground. Scores of native animals—likely including finches and short-eared owls—will be captured and held in aviaries both on and off the island. These aviaries have been built, and they've been home to some captive finches as a trial run. Campbell expects it will take 10 years, $26 million, and 35 shipping containers full of poisoned cereal to clear Floreana of rats. "Rodent eradication requires getting in every habitat," Campbell said. "You can't exclude any area. It has to be 100 percent or you fail."

Afterward, the island will have to inspect every incoming ship carefully forever, to prevent new rats from coming ashore. The long-term success of the plan relies on the Ecuadorian government following through with intensive biocontrol in perpetuity. They already have a biocontrol station. When I arrived, I checked in there and watched as two young men pawed through my suitcase in a desultory fashion.

Ecology is complex, even on small islands, and things don't always go according to plan. In 2012, for instance, Campbell helped round up the 60 Galápagos hawks that lived on Pinzón Island, a steep volcanic nubbin in the Galápagos chain, so they wouldn't get sick from eating the rats that Campbell was about to poison to save the island's "tiny and very soft" tortoise hatchlings. But when the rare raptors were released back into the wild after a couple of weeks, they began dropping like flies. It turned out the poison was still lurking in lava lizards—and the hawks were preying on the lizards and getting poisoned that way. "I was getting beaten up by pretty much fucking everybody," Campbell says. Just a dozen of the birds nest there now. But Campbell pointed out that baby tortoises have been born to the ancient tortoises that live there—the first to survive in more than 150 years. If a small percentage of native animals die, that's fine with him, because that's better than 100 percent going extinct.

Project Isabela was widely hailed as a success. With the goats gone, endangered giant daisy trees grew back—but on Santiago Island so did huge, Sleeping-Beauty-esque tangles of non-native blackberry. Removing non-natives doesn't always magically and completely restore an island, especially when other introduced species have made it home as well. Though the project was lauded, there are some who feel the organization should have budgeted an extra million for blackberry removal.

"It continues to unveil itself," Campbell said about Santiago, with a shrug. "It is unrealistic to put a system back 500 years." For Campbell the goal isn't time travel, it is stopping extinctions.

Perhaps because of his disdain for comfort, Campbell has thrived in the harsh volcanic landscape of the Galápagos, with its strange and wonderful wildlife. He married an Ecuadorian jewelry designer, and they have a daughter. But Campbell is frustrated with the slow pace of the work. There are thousands of islands out there—and so many of them are in the midst of an ecological crisis. "We are barely scratching the surface," he said. "I will never, in my lifetime, run out of a job."

His job might change, though. While most people are generally cool with killing mice and rats, many are squeamish about killing them in such a painful way. And the acceptability of killing rodents at all is potentially declining. Recently, 25,000 Parisians signed a petition to stop the "genocide" of rats in the City of Lights. In addition, the mass-poisoning eradication approach has logistical limits—and if he's already approaching those limits

California condor. SUSAN HAIG, U.S. GEOLOGICAL SURVEY

Protestors outside the L.A. Zoo on November 30, 1986. Many objected to the plan to take every wild California condor captive in order to breed them and save the species. Protestors felt the plan would sacrifice the birds' wildness. MIKE SERGIEFF, PHOTO COLLECTION/LOS ANGELES PUBLIC LIBRARY

Nonhuman Rights Project founder Steven Wise with a client, Teco the bonobo, in 2016.
PENNEBAKER HEGEDUS FILMS

Washington State biologists move Wolf 47 to a safe release site after putting a collar on her.
PHOTO BY RICH LANDERS. *THE SPOKESMAN-RE-VIEW*

Joy Adamson with the three lion cubs her husband inadvertently orphaned. One of the cubs, Elsa, would grow to adulthood at the Adamsons' home. ELSA CONSERVATION TRUST

Chai leaving Seattle in 1999 to visit a zoo in Missouri, where zookeepers hoped she would mate with a bull and become pregnant. Her baby, Hansa, was born in 2000. ERIC SCIGLIANO

Mary the elephant was hanged for murder in 1916. THOMAS G. BURTON-AMBROSE N. MANNING COLLECTION, ARCHIVES OF APPALACHIA, EAST TENNESSEE STATE UNIVERSITY, JOHNSON CITY, TENNESSEE

An 'akikiki on Kaua'i. Fewer than 500 exist today. CARTER ATKINSON, U.S. GEOLOGICAL SURVEY

Yoina, a Matsigenka girl, in the Yomibato River with her pet saddleback tamarin. CHARLIE HAMILTON JAMES

Paul Ward, founder of Polhill Protectors, with a rat trap in Wellington, New Zealand. COURTESY OF THE AUTHOR

The toutouwai or North Island robin has little fear of humans. It cannot co-exist with rats, which eat birds and eggs. WOLFGANG KAEHLER/GETTY IMAGES

Kris MacDonald, then CEO of the Ngātiwai Trust Board, posing with a carving of an ancestor at the Waitangi Treaty Grounds. His tribe see themselves as guardians of the kiore—the Pacific rat. COURTESY OF THE AUTHOR

The "death row dingoes" were used to kill goats on an island in Australia. The plan was to kill the dingoes after the goats were gone. They were fitted with GPS collars and—in case they could not be located to be shot—"a small, delayed-release toxic implant was also inserted under the skin between the shoulder blades of each dingo." BENJAMIN ALLEN, UNIVERSITY OF SOUTHERN QUEENSLAND

A giant tortoise at the Asilo de la Paz refuge on Floreana in the Galápagos. Returning tortoises to "the wild" on Floreana would likely require killing all the island's rats. COURTESY OF THE AUTHOR

Farmer Claudio Cruz with some of his cattle on the Galápagos island of Floreana. Livestock would have to be kept indoors for weeks to months during any effort to eradicate Floreana's rats with an island-wide drop of poison. COURTESY OF THE AUTHOR

Arian Wallach, a "compassionate conservationist," in the field in Australia. She believes killing to save species is unethical. COURTESY OF THE AUTHOR

Mike Letnic, Dan Blumstein, and Katherine Moseby search for a radio-collared cat at Arid Recovery Reserve. They are using cats to help native animals evolve defenses against these introduced predators. COURTESY OF THE AUTHOR

A greater bilby. These Australian mammals cannot coexist with feral cats. JASMINE VINK

The Gundestrup cauldron, featuring the antlered god Cernunnos. On display at the National Museum (Nationalmuseet) in Denmark. MALENE THYSSEN, WIKIMEDIA COMMONS

Orlando Yassene with a wild greater honeyguide in northern Mozambique. Humans and honeyguides work together. The bird locates a beehive; the human opens the hive; the humans eat the honey and the birds eat the wax. CLAIRE SPOTTISWOODE

on an island with just 100 humans, it looks like a nonstarter for larger, more populous islands.

————

Campbell and Island Conservation focus on introduced species on islands. But some conservation killing happens on continents—and the species that are killed aren't always from across the ocean. Sea lions that eat endangered salmon on the West Coast are trapped and given lethal injections. To save threatened herds of caribou, the government of Alberta has shot more than 1,000 wolves from helicopters.

To protect some rare songbirds, conservationists kill brown-headed cowbirds. The cowbird is a nest parasite: It lays its eggs in the nests of other bird species, then takes off and repeats the trick, laying as many as three dozen eggs in various borrowed nests. When these cowbird chicks are born, usually before the biological chicks of the nest's builders, they grow quickly and out-compete the parent's real babies, and sometimes even heave them out of the nest. Cowbirds evolved this trick back when they followed huge bison herds across the plains, feeding on goodies in their dung. When those herds were systematically hunted out in the 1870s as part of an unofficial military strategy against the continent's Indigenous people, the cowbirds took up with domestic cattle, and fanned out across the continent to every place cattle were—which is to say nearly everywhere. So conservationists protecting the Kirtland's warbler in Michigan's jack pine forests trapped and killed cowbirds there every year for four decades.

In the United States, barred owls moving west on their own are shot because they compete for nesting sites with the threatened spotted owl. Here, the birds being shot are so closely related to spotted owls that they can mate and have fertile offspring. And those hybrid offspring might arguably be better adapted to the current state of the Pacific Northwest. Spotted owls require old growth habitat, eat a smaller set of prey items, and are shier and less aggressive. Barred owls can live more places, eat more things, defend their territory better. Whether you see the interaction between the cousins as an "invasion" of foreign owls that will "contaminate" the spotted owl genome and drive them extinct or whether you see it as hybridization creating adaptive diversity in a rapidly changing environment depends on your values. In the meantime, federal officials have shot some 3,135 owls.

The barred owl case brings up an uncomfortable fact: Many people see species as having final value, but the borders between species can be fuzzy because the term "species" itself has several possible meanings. Some use the "biological species concept," which says that if any two individuals can mate and produce fertile offspring, then they are the same species. Some take an ecological approach, classifying together organisms that act the same way and do the same things in the environment. Some look at the physical details of the organism, its colors, shapes, number of spots. And some simply run the genome through a computer and assign it to a species based on genetic similarity. There's no one right answer to the question of how to define species; it really depends on what kind of questions you are asking about them. Ultimately, "species" is a human concept rather than a biological reality.

If you ask whether a California condor and a Floreana mockingbird are different species, any of these definitions would tell you they are distinct. But whether mockingbirds from various Galápagos islands should be considered separate species or merely subspecies is a scientific debate that has continued since Darwin's day.

Complicating matters is the fact that many lineages that split off from one another re-encounter each other later, hybridizing and coming back together. The tendency of plants to "naturally" hybridize has been well-known for generations. Hybridization in animals was for a long time considered rare. When biologists found hybrids, their first hypothesis was often that the mixing had to have been caused by humans, and hybridization was automatically assumed to be a threat to species integrity. Hybrid animals were also generally thought to be "less fit" and thus to jeopardize the survival of the species. But recent work has shown that hybrids are common. Some 10–30 percent of multicellular animal and plant species hybridize regularly, according to evolutionary biologists. Even in the animal kingdom, related species often re-encounter one another after a period of separation of millennia or more, and then interbreed. And their offspring are not always less fit. Indeed, an influx of genes at the right moment can increase fitness and *prevent* extinctions—as seems to be the case when California tiger salamanders got a dose of new genes from barred tiger salamanders. A new pool of genetic diversity to draw from can help a lineage adapt to changing conditions. This has even happened in the family tree of Galápagos mockingbirds. The Genovesa mockingbird looks to be a sister species of the Española and San Cristóbal mockingbirds, but with some extra genes it picked up when it mated with mockingbirds from Isabela and surrounding islands.

One approach to sorting out whether animal hybrids are "good" or "bad" in terms of biodiversity is to ask whether the resulting organisms will be more resilient and likely to persist in the face of the ongoing processes of environmental change that we humans have kicked off. Another way is to investigate whether the individual hybrids themselves will be more or less able to be happy and flourish. Every case is different, but without being too flip about it, I tend to think that if two populations create hybrids that are more fit, then the two parent populations were probably pretty closely related to begin with and so their intermixing is not very high on my list of environmental problems to worry about. Put another way, I'm willing to accept old growth forests filled with "sparred owls" because it's better than no owls at all.

We ourselves are hybrids. Within the last million years or so, many of our ancestors mated with at least two other species: the Neanderthals and the Denisovans. Those of us with Oceanic and Asian heritage might be as much as 7 percent Denisovan. Those of us with mostly non-African heritage are between 1–4 percent Neanderthal, and at about 20 percent of the total Neanderthal genome is floating around in the human gene pool somewhere. Some of these genes are for hair and skin traits, suggesting that these hybridization events helped populations of *Homo sapiens* adapt to new climates and altitudes they encountered as they migrated northwards out of Africa. Genes from these other species may have also helped protect us against new diseases.

In addition, most cells in animal, plant, and fungal bodies include mitochondria—a structure enclosed in a membrane that converts broken-down bits of sugars into adenosine triphosphate (ATP), which carry energy throughout our bodies to power everything we do. Mitochondria used to be an independent, free-living bacteria, until it was absorbed by one of our very distant ancestors and took up residence inside our cells. To this day, mitochondria have their own DNA, separate from our main packet of chromosomes. When we reproduce, the mitochondria reproduce in parallel. Thus even *individual* organisms are amalgams of at least two individuals working as a unit. (Lichen, similarly, is actually a partnership between an algae or a cyanobacteria and a fungus acting as a single organism.)

On top of that, animal bodies host bacteria, fungi, protozoa and viruses, whose cells outnumber our own by a large margin. Many of these organisms are symbiotic. We provide them with an environment and nutrition and they help us digest food, fight infection, and more. An animal completely stripped of their microbiome is a sick animal. If the dividing lines between species are blurry, so too are the dividing lines between individuals.

More recently, microbiologists have studied "horizontal gene transfer" (HGT) in microbes, where genes from one species pass directly into the genome of another species. This is most common in bacteria and archaea, but plants are known to have dabbled, too. Ferns somehow copied a gene from a mosslike species of hornwort which allowed them to sense low levels of light in deep shade, helping them thrive in the dark, moist environments so many ferns occupy today. And animals have collected genes from other species as well. One 2015 study took advantage of the boom in whole-genome sequencing to look for bacterial genes in the DNA of 10 primates, 12 flies, and 4 nematodes; genes likely acquired from bacteria were found in all of them. The authors concluded, "Far from being a rare occurrence, HGT has contributed to the evolution of many, perhaps all, animals and that the process is ongoing in most lineages. Between tens and hundreds of foreign genes are expressed in all the animals we surveyed, including humans."

Together, compound organisms, horizontal gene transfer, and hybridization create a picture of evolution operating not like as a branching tree with a perfect, unchanging fruit on the end of each branch, but as fungal network, with genes flowing sideways between lineages as well as "verti- cally" from parents to offspring. Species drift apart, then back together; they dead-end often. Sometimes one swallows another whole. It is less like a tree and more like the complex networks in forest soils, a dense tangle of myce- lium and roots and bacteria, splitting and merging and growing into one another. And the species we see today are not the final products. On the contrary, we live in one moment in somewhere near the middle of a story billions of years in the telling. When we are tempted to stop lineages from changing or stop species from hybridizing, we must ask ourselves: are we really preserving biodiversity with these actions, or are we thwarting it?

———

After visiting the farm on Floreana, Campbell and I hit the beach by my hotel, white sand bordered by black volcanic rocks spangled with crimson Sally lightfoot crabs. Offshore, sea turtles popped their heads above the waves. Opuntia cactus were outlined against a cloudless sky. I saw Eden. Campbell saw trouble. He pointed at an almost imperceptible depression in the sand, about the size of a silver dollar. It was the footprint of a cat, and it was less than a meter away from the unprotected ground-level burrow of an endangered Galápagos petrel. Campbell and his team are also working on a cat sterilization campaign. "I know the name of every pet cat on this island," Campbell said with grim determination. "There's one fertile cat left."

Campbell had some more work to do, so I decided to go snorkeling, even though the water was pretty cold. Most tourism in the Galápagos is through organized cruises, and none of them were on the beach at the moment, so I had the whole place to myself. The water was just a bit murky, which I knew meant that it was filled with nutrients. I watched sea turtles feed with joy. A juvenile sea lion came to say hello. I backed up, trying to keep my distance, like a good tourist, but he or she was having none of it. The sea lion turned playful circles around me in the water and looked me right in the face. I felt a soft, slick body against my stomach then my back. I was thrilled but felt guilty about "letting" the animal come so close—although I'm so clumsy in the water all I could really do is stay as still as possible. I found myself thinking, "You wouldn't be so friendly if you knew what monsters we humans are"—the same old misanthropic environmental narrative I absorbed as a kid. Humans are the virus, etc.

I swam to shore and sat on the rubble of a million broken and weathered sea urchin spines, collecting my thoughts, shivering. Here on Floreana, humans are trying to undo the damage they have caused, to repopulate the island with the species (or subspecies) that are missing. Their intentions are good. Maybe some of those involved are motivated by the idea that human influence must be cut out of ecosystems like a melanoma, but Campbell seems simply to act from an impulse to fight extinction, preserve the diversity of life. And if things go as planned, all the rats will die at once during the poisoning, and no one will have to kill any more in the future. The killing can stop. Claudio Cruz won't have to spend a fortune on poison to protect his yucca. Rats won't die on the edges of his fields every year. Baby tortoises will hatch and grow old. Long after I'm dead, they could walk this island, taking their unhurried steps, nibbling opuntia cactus fruit. Maybe it would all be worth it.

12

The Friendly Toutouwai

When I visited Campbell, he was still in the planning stages of the Floreana eradication. I wanted to see what it was like when the killing started. So I went to another Pacific Island that struggles with introduced predators: Aotearoa, also known as New Zealand. Campbell had told me that Kiwis are the world's experts in killing non-native predators.

Aotearoa broke off from the mega-continent known as Gondwanaland about 85 million years ago—before the age of mammals—and has been isolated ever since. Before humans arrived, there were no four-footed mammals at all—just a few bats, seals, and sea lions—animals that could make it to the islands on their own. Many ecological roles filled by mammals in other places were here filled by birds. Five-hundred-pound flightless Moa were like deer or gazelle, grazing and browsing. They were hunted not by wolves or lions but by massive hawks and eagles. Haast's eagles had an eight-and-a-half-foot wingspan and talons big enough to comfortably grasp a human head. Roles taken by rodents in other places were here filled by giant flightless crickets called wētā, some as big as adult mice. (Weta Digital, the special effects company co-founded by Peter Jackson, of *Lord of the Rings* fame, is named after these admirable beasts.)

Aotearoa was the last large landmass on Earth to be discovered by humans, who arrived in a fleet of oceangoing canoes launched from the

Society Islands less than 1,000 years ago. These people, who became known as the Māori, brought animals with them, including at minimum dogs and kiore—also known as the Pacific Rat. The kiore was likely not a stowaway. On the contrary, the Māori brought it along on purpose. They let the rats run wild in the forest, then trapped them in the winter, when they were most plump. They could be preserved by being packed in their own fat—a kind of rat confit. The Māori made cloaks from their soft, beautiful fur.

There was a pulse of extinctions when people came. Humans killed off the enormous moa, along with other species or waterfowl and rails—all hunted for food. Haast's eagles went—starved out when their Moa prey disappeared. Kiore can be predators of land snails, wētā, lizards, and burrowing birds such as petrels and shearwaters, so they may have caused some extinctions of their own as well.

When Europeans arrived in 18th century, they began introducing even more species to the mix. Some were brought to be hunted, including the European rabbit, the brown hare, many species of wallaby, and a grab bag of deer from all over the world. Others were brought as farm animals: pigs, horses, cattle—and above all, sheep. The stoat, ferret, and hedgehog were brought in order to eat pests, but they ate native birds and insects too. Cats came as pets. Brushtail possums were introduced from Australia for their undeniably luxe fur and thrived on a diet of native New Zealand plants and perhaps the occasional insect or bird egg. And some were unwanted tagalongs: the Norway and ship rats, the house mouse. When cats and rats and stoats showed up in a land of flightless birds, untended nests filled with eggs right on the ground, and animals whose main defense against flying predators was to *freeze*, it was like an all-you-can-eat buffet.

Since European arrival, some 27 species have gone extinct in New Zealand, according to the IUCN red list. Of these, 15, by my count, are known or suspected to have been casualties of introduced animals. The first was *Hoplodactylus delcourti*, an enormous gecko—37 centimeters long, *not including the tail*—known from a single specimen found stuffed in a French museum in the 1980s and probably collected between 1833 and 1869. Even though just one specimen exists, experts have some thoughts on why it went extinct. The IUCN listing reads, "The causes of this species' apparent extinction are unknown, but if it genuinely occurred on New Zealand's North Island the destruction of primary forest following European colonization, and the introduction of mammalian predators including cats, black rats and feral pigs may have been causal factors."

The most recent extinction was in 1972, when the last ground-nesting mātuhituhi, or bushwren, died in an unsuccessful last-ditch semi-captive breeding project on Kaimohu Island.

To tell the truth, I was surprised that "only" 15 species have been driven extinct by introduced predators in Aotearoa since the Europeans came. To be fair, the real number is probably higher. Many species likely disappeared without leaving a record. And those 15 may not be the last. The IUCN lists more than 150 New Zealand species as threatened at least in part by "Invasive and other problematic species, genes & diseases." Those in danger include such charismatic species as the kakapo, a flightless green parrot full of goofy charm, and five species of kiwi, the flightless needle-beaked fluff balls that are a national symbol. Many of this group only persist on small islands without introduced predators; they are functionally extinct on the "mainland" comprising the North and South Islands of Aotearoa.

Having lost 53 species of birds since human arrival, New Zealanders don't want any more of their threatened species to disappear, and they are willing to get their hands dirty to prevent it. "Predator Free 2050" is an ambitious government program to kill all the rats, stoats, and possums in the entire country. (You will note that they have chosen to focus on just a few species. Feral cats are also a threat to many native species, but because 41 percent of New Zealanders own a pet cat, the politics of killing them are considerably different. Pigs also cause a great deal of destruction to native plants, by rooting them up, but there's a large group of pig hunters who do not want them removed.)

Once the predators are eradicated, rigorous biosecurity will be enforced to keep the unwanted mammals from returning. This watchful guarding will presumably have to continue for millennia—and rodents like rats are hard to intercept. I asked James Russell, a University of Auckland ecologist and a proponent of Predator Free 2050, how confident he was that humans would still be on guard against mammal predators 1,000 years from now. He replied, "I honestly have no idea, but I can only do my best and strive for what I believe is the correct course of action today, and hope those that follow share my values."

Eco-minded citizens all over New Zealand have joined the Predator Free fight. School children are taught how to set and empty traps. Community groups, Indigenous iwi (the Māori equivalent of "tribe"), and other teams undertake predator killing and share images of their trophies on Facebook. Government websites offer advice for families who want to trap in their backyards. One features a short video that opens with a cheerful host holding a stiff, dead rat, to which he says, "You didn't see that one coming, did you!"

In one well-publicized incident, a possum hunting contest was held by a school as a fundraiser. Whoever killed the most possums in three nights— measured by weight—won. At another school, carcasses of the dead possums killed in a hunt were dressed up in costumes: a bride, a sunbather, a skinned possum in boxing shorts and gloves. "If you see a possum on the road while driving, the general rule is to run it over instead of slowing down or trying to avoid them," the New Zealand Pocket Guide advises.

The Department of Conservation (DOC) routinely air drops green nubbins of bait laced with a poison called 1080 over large areas of forest to kill possums and rats. Forest & Bird, a conservation organization, explains on its website why the poison is used: "One aerial application can kill over 95 percent of possums and close to 100 percent of rats in the targeted area, although rat numbers can bounce back in one or two years. However, birds and other native species can benefit greatly from having one or two good breeding seasons without large-scale predation by rats. 1080 can also kill all or most stoats after they feed on the bodies of rats that have been killed by 1080." In other words, with no realistic prospect of eradicating possums and rats anytime soon, conservationists are settling for an endless cycle of killing.

Derived from a compound found in Australian plants to keep mammals from eating them, 1080 is considered to be more humane than brodifacoum, primarily because it kills animals more quickly—in hours instead of days or weeks—and the associated suffering is rated as merely "severe" instead of "extreme." But let's be real. It is not a nice way to die. In a lab experiment, researchers fed carrots dosed with 1080 to eight possums. It took them anywhere from four to 18 hours to die. About three hours in, they lost physical coordination and displayed "sunken, glazed eyes and lowered ears." They then retched and vomited and "spent most of the time until death lying, showing spasms and tremors. Five showed seizures while lying prostrate." The Royal New Zealand Society for the Prevention of Cruelty to Animals says that "these substances cause such intense and prolonged suffering to animals that we believe their use can never be justified."

Scientists at the Toi Ohomai Institute of Technology in Rotorua are studying the possibility of using drones to drop 1080 poison to kill possums. Drones would make it easier to distribute poison where there are no trails. They could also protect conservation workers from the emotional toll of killing, according to drone researcher Craig Morley. "You do this work because you are trying to protect the birds and the plants but you end up killing things," he told me. "The first time you start doing that, it hurts you. It is not nice." The drones could put that hurt at a remove. "Someone did compare it to war," Morley says. "You used to be able to see your opponent. Now you just press a button and you fire a missile, You become a little bit detached from the reality that you have killed something or somebody over there."

Morely sees this emotional distancing as a good thing, because he thinks the killing is absolutely justified. "We are losing about 26 million native birds a year. We've got one bird that is on the brink of extinction. We've got to get cracking with this. Otherwise New Zealand is going to become very quiet. And that's sad."

Kiwis love animals. In February 2020, a woman from Pukekohe who didn't take proper care of her pet Netherland Dwarf rabbit, Lambo, was sentenced to 90 hours of community service, banned from owning any new pets for ten years, and fined NZ$521.40. And yet two years earlier and a half-hour drive east, the government dropped 1080 on 22,000 hectares of forest in and around Hunua Ranges Regional Park, killing untold numbers of rats and possums with very similar capacities for pleasure and pain. Their deaths were justified to protect the North Island kōkako, a gray bird with a black mask, a violent waddle, and a soulful minor-key song. There are about 1,400 breeding pairs of North Island kōkako in Aotearoa, up from less than 800 pairs in 2008. The bloody work of killing their predators may yet save them.

After I arrived in the country, I began to notice symptoms of the national obsession with killing introduced species. At a museum gift shop in Auckland, I found a picture book for children called *It's my egg (and you can't have it)*, which turned out to be about how great it is to trap and kill stoats because they eat kiwi eggs. It featured a very evil-looking stoat and a two-page spread that just had the word "WHACK" written across it in blood-red letters.

I headed to Kapiti Island (the full Māori name is Ko te Waewae Kapiti o Tara Raua ko Rangitane), about an hour north of the city of Wellington. The island's location at the boundary of Māori territories made it a strategic military location in the early 19th century. Later, it was used by whalers to render blubber and then by European farmers. The New Zealand government bought much of the island in 1897 explicitly for conservation, and the first flightless birds were moved there to try to save them from extinction in the 1890s and 1910s. The work of getting rid of introduced animals started early too. Two thousand feral goats were shot in 1928. Cats, deer, sheep, cattle, pigs, and dogs followed. Possums weren't tackled until the 1980s, when conservationists spent six years killing 22,500 of them with 1080 poison, traps, and dogs. Some 50,000 or so kiore and Norway rats were killed off in the 1990s with aerial application of brodifacoum. By 1996, the island was finally ready to live up to the government tourism agency's marketing slogan for the whole country, launched in 1999 and still in use today: "100% Pure New Zealand."

Kapiti is about six miles long and a mile wide, and a boat ferries a maximum of 160 tourists a day there to see rare species like the stitchbird, or hihi, a busy songbird, and the weka, a flightless rail that stalks slugs and grubs like a miniature Tyrannosaurus. After a short crossing, my boatload of tourists got an orientation chat from a guide named Rochelle, who told us the history of the island. Apparently, in 2010, someone saw a single stoat on Kapiti, kicking off a massive search for the animal and any of its compatriots. "It cost one million dollars to catch," the guide told us. "No one knows how it got here."

After we were briefed, we were free to explore on our own. I headed up toward the center of the island into the dense and damp forest. The flora of the island was a mix of forms that read as tropical to my eye—like the elegant, upswept nikau palm—and those that read as temperate, like the silver fern. But my botanical bias was showing. I stopped and closed my eyes, listening for fauna. The forest was alive with birdsong. Chatty saddlebacks in pairs called out their Māori name: "Tīeke, tīeke, tīeke." I heard a burbling, lubricated-sounding call. I opened my eyes and located the source, a blue-and-brown bird with a striking white cravat of feathers, the tūī.

Farther up the trail, I encountered a series of feeders for the hihi. There are some 3,000 of these birds left; they are extinct on the mainland, probably because of a combination of rats and loss of forest. At first, the birds were too fast for me to get a sense of. They buzzed back and forth between the trees and the feeders with the speed and purpose of bees. I sat down nearby and listened to the busy whirr of their feathers. When one finally paused for half a second, I made out a round songbird with a black head and white spot behind each eye—a bit like a killer whale.

On the way downhill, a lime green red-crowned New Zealand parakeet, or kākāriki, barreled down the trail, its wings fixed in gliding mode. It seemed absolutely unconcerned at my presence. The birds were clearly in charge here, the humans merely their servants and worshipers. I noticed a brown boxlike structure on the side of the trail. It was a rat trap baited with poison. With boats of tourists docking each day, the caretakers of Kapiti must remain ever vigilant.

Emerging from the forest I found a moist meadow through which strode an imperious looking weka—fat and brown and flightless. I spotted a staff member and asked him about the birds. He told me they were introduced in 1896, back when people were less particular about keeping subspecies apart, so they are a mix of North Island Weka and South Island Weka. A 2016 Kapiti Island ecological restoration strategy lists the island's weka's conservation status as "N/A due to mixed provenance." They were also not known to be on Kapiti at any time in the past. And, he added, they can be a bit of a pain. As predators, they will eat other endangered species. These weka were kept fenced in one part of the island to keep them from eating eggs or chicks at a sooty shearwater colony nearby. (One of the few advantages of protecting flightless birds is that you can keep them where you want with fences.) Finally, he added, the weka had learned how to open ziplock bags to get at the packed lunches of sandwich-bearing tourists. Nevertheless, there was no real talk of removing the weka. "They are not doing well on the mainland," he said.

The more you look into the details, the less an island like Kapiti seems like a trip back in time and the more it seems like a glimpse at a possible future: species in new places, hybrids at least provisionally accepted, and humans using unobtrusive management techniques like fences to allow many species to coexist in a small area. The good news is that if Kapiti is any indication, such management can create places filled with life where humans can feel both inspired and satisfyingly incidental. It would do us all good to be ignored by a kākāriki once and awhile.

Next I visited Zealandia, a 225 hectare sanctuary in urban Wellington, cleared of introduced predators with traps and poisons and maintained by enthusiastic Kiwis and a very serious fence to keep out non-native predators. As you move through the fence, you feel a bit as if you are entering a prison, but inside it is a lovely place. I saw grave and resplendent tuataras, ancient reptiles with a crest running down their backs. Rats like to eat their eggs and babies outside the fence. I spotted a South Island takahē, an extremely sturdy flightless bird that was native to the other main island of New Zealand. (The North Island, where Wellington is, lost its takahē in 1894.) It is deep blue, with a tomato-colored beak that extends up onto its forehead. I spent a lot of time trying and failing to get a good picture. There are fewer than 300 of these birds in the world.

I almost stepped on a tiny, big-eyed bird sauntering down the path on comically long legs. It was a toutouwai, also known as the New Zealand robin. Like birds called robins in other countries, this bird is not only fearless, it seems actively interested in people, probably because of the invertebrates we kick up. My tour guide, Danielle Shanahan, Director of the Centre for People and Nature at Zealandia, showed me a trick. She kicked over a bit of dirt, then took a step back. The toutouwai glanced at us for a moment, weighing the risk, then walked up to the freshly turned earth to investigate, looking for bugs to eat. I absolutely surrendered. "This is my new favorite bird," I said.

On the way out, we stopped by Zealandia's gift shop, which sells throw pillows with pictures of non-native animals with large red Xs embroidered over them—conservation killing as home decor.

I asked Shanahan about the vilification of the introduced animals. Does she ever worry about the way it shades into ascribing evil motives to animals who, in reality, are just living their lives, completely unaware they are "invaders"?

"It is very easy to get caught up in just getting rid of stuff," she admitted. That's why, she said, she prefers conceptualizing the goal of ecological restoration in New Zealand as making the country "nature rich" instead of "predator-free."

Eleven thousand school kids a year come to Zealandia and learn about the environmental history of their home. And the native animals are doing so well inside its boundaries that they are beginning to spill over into the neighborhoods around the park. I saw a tūī, the striking bird with two white puffs of feathers at its throat, singing lustily in the parking lot outside the fence.

This "spillover" effect would be limited without the work of volunteer groups who trap and poison rats and other introduced predators in forested areas outside the fence. One such group is the "Polhill Protectors," 65 volunteers who patrol a 75 hectare slice of urban greenbelt with the zeal of an army. Volunteer wrangler and "lifelong bird nerd" Paul Ward took me on a tour of the traps. Ward started running in this slightly scruffy bit of forested land for exercise. But then one day in 2006, mid-jog, he saw a huge, rust-colored kākā parrot in a tree. "I'd never seen one anywhere other than an offshore island," he said as we parked his car and plunged into the trees. "I was just gobsmacked."

These hills are by no means pristine wilderness—a good chunk of the plants here are introduced—but they are green and shady, and the birds seem to be able to live here, as long as humans continuously work to keep the density of predators down. "To see these birds back in a lived environment is really cool," Ward said.

We stopped at the first trap—a long wooden box with a rat-sized entrance on one end, stamped with the logo of the Polhill Protectors: a Māori bird person with outstretched wings. The mechanism of death was inside. Ward slipped an 8 mm ratchet spanner out of his pocket and deftly unscrewed the end of the trap. This one was empty. "The rat steps on the platform and it dies instantly." He noted that the traps have been okayed by the National Animal Welfare Committee.

Ward's next project, dubbed "Capital Kiwi," seeks to create a safe space for the iconic flightless bird in Wellington by operating the largest community owned stoat trap network in New Zealand. As of 2020, they maintain 4,443 traps over 23,455 hectares.

Ward isn't Māori himself, but he said he is inspired by the Māori concept of kaitiakitanga. Meaning "guardianship," it holds that individuals or groups bear duties of care to specific taonga, or "treasures"—particular species, places, items, or stories with which the humans have a relationship. "Trapping is part of being a citizen," Ward said. His community evidently approves of the job he is doing. In 2018, Ward was voted Wellingtonian of the Year in the environment category.

We checked several more traps, all empty. Apparently either the rats were wising up, or the Protectors had knocked them back so far that there were none left to kill. As we emerged from the woods to return to Ward's car, I noticed a soccer backstop in a sports field that had been painted with cartoonish versions of several native and introduced species. Each had a point value painted on its body. The idea, Ward explained, is to kick your soccer ball so it smashes the invaders. But if you miss and hit a native, you lose points. I asked if the game isn't a bit aggressive. "Those animals are massacring native species," Ward said. "The choice is not to do anything and lose our native taonga or do something and have them in our backyard."

Ecologist Chris Thomas says the native species of Aotearoa are clearly cultural treasures for the humans there, but keeping some species going—especially flightless species like the takahē and the kiwi—may be a forever fight. These species evolved to survive on isolated islands without many mammals or ground predators, but now thanks to human beings, Aotearoa are not isolated islands. The country is tightly woven into a global web of trade, tourism, and other travel. Thus the distinct traits of its island-adapted species are no longer adaptive. According to Thomas, they are evolutionary dead-ends. Their loss hurts, but as new species arrive, they will adapt and drift and become unique species that will replace them. Over the very long term, he says, biodiversity will bounce back.

Thomas contrasts two birds: the big flightless takahē, which evolved from swamphens of the genus *Porphyrio* that arrived in New Zealand about 2.5 million years ago; and the smaller flying pukeko, which also evolved from newly arriving *Porphyrio* swamphens, but just 500 to 1,000 years ago. Takahē can't cope with predatory mammals; pukeko can. "The takahe represents New Zealand's past, and the pukeko its future," Thomas writes. Large flightless birds with low reproductive rates are going to require a lot of help if we want to keep them around. In general, island forms simply aren't well adapted in a globalized world. We must decide if we want to try to preserve them anyway, and if so, at what price.

Thomas says that he's not overly concerned about conservationists killing animals in their work. "In the end, each of us is born and lives and dies," he says. But he's not a full-throated supporter of Predator Free 2050 either. "I question whether it is worth, at the moment, extending the strategy to the entire land surface of the country," he says. "That's a lot of rat death with a modest chance of success." Then again, the rats cause plenty of suffering too. "The rat doesn't want to die of poisoning; the kiwi doesn't want to be attacked. I don't believe that there can be an ethically correct answer,"

Thomas says. "Nearly all species that have ever lived on the planet are gone. Mostly all sentient organisms are dead. There is a minuscule fraction alive today. We can trade off the deaths of one thing versus the deaths or survival of other things. But there can't be an ultimate right or wrong."

Thomas is based in the UK. At first it seems like everyone in New Zealand has fully embraced the Predator Free 2050 campaign. But then I begin to meet a few people with more heterodox views. Among them is Wayne Linklater, who was then a professor at Victoria University in Wellington, and Jamie Steer, a senior advisor in the Biodiversity Department at Greater Wellington Regional Council. In 2018, they published a critique of the campaign in the journal *Conservation Letters*, calling it "unachievable" and "flawed."

"It's the ends justify the means in extreme," Linklater says. "How can we foster a caring relationship with nature among future generations if our tools are suffering? I think we are on the wrong side of history on that one."

Their objections aren't solely on the grounds that the means of achieving the goal involve animal suffering. They say that while the issue of non-native predators got all the attention, the clearing of native ecosystems for human use is an overlooked threat. Two-thirds of Aotearoa's forests are gone. Almost all the country's lowland forests and grasslands are history. Ninety percent of its wetlands have been drained. Surely these changes also have something to do with the decline of many native species. "[I]f biodiversity recovery is our ultimate goal then predator eradication is secondary to the need for somewhere for biodiversity to live," they write. Preserving and resorting habitat can go beyond just declaring an area a park. Linklater suggests that "refugia" for endangered reptiles can be created—like rock piles where they can hide from mice and rats. Native animals could also be given supplementary food so they could reproduce more, possibly staying ahead of the introduced predators.

Presumably, some of this work could be paid for with the money currently going into Predator Free 2050. Actually achieving the vision will cost a huge amount of money. One estimate suggested it might cost NZ$32 billion, all told. Every dollar spent killing can't be used restoring habitat—or on other national priorities, like reducing poverty and making housing more affordable.

Linklater and Steer also worry that simply removing rats, stoats, and possums without doing anything about mice, cats, deer, pigs, rabbits, and all the other introduced species risks unpredictable and potentially catastrophic ecological consequences—think of those blackberry vines on Santiago Island.

In addition, they say, not everyone in New Zealand is okay with strewing poison all over the land, or even supportive of the goals of the project. "The eradication of some introduced species is also contentious because some Māori regard them as culturally important," they write. "The Pacific rat, for example, while targeted by Predator Free 2050, is protected on some Māori lands."

Rats? Protected? This I wanted to see for myself.

———

The next thing I knew, I was on board a sailboat with Kris MacDonald, the genial CEO of the Ngātiwai Trust Board. We were en route to meet up with a natural resources field team from the iwi who were surveying for kiore on the remote island of Mauitaha, which is, to my knowledge, the world's only rat preserve. The day was fine and as we followed the coast, MacDonald told me stories about the various rocks and peaks, most of whom used to be people. "That's our tūpuna right there," MacDonald says, using the Māori for "ancestors." The fact that their lineage can be traced back to these rocks actually forms a part of the legal case the Ngātiwai are making for control over their traditional territory, he says.

In what is now called the United States, the federal government signed over 300 treaties with Native nations, most of which the U.S. government proceeded to break or ignore. In Aotearoa, in 1840, British colonists signed a single overarching treaty with chiefs, or rangatira, representing most Māori iwi. The Treaty of Waitangi was written out in English and Māori and the two versions were ever so slightly different, with the Māori version, perhaps unsurprisingly, reading as more favorable to the iwi. In the English version, the iwi gave up their sovereignty in exchange for full citizenship and retained their physical property. In the Māori version, the iwi gave up just the "kawanatanga" of their land, which is more like "governance," and retained "'tino rangatiratanga' (full authority) over 'taonga' (treasures, which may be intangible)."

After a protest movement that began in the 1970s, a body called the Waitangi Tribunal formed to investigate whether the spirit of the treaty had been kept. Many iwi, including the Ngātiwai, have filed claims for land, fishing rights, and rights to access and manage lands and marine areas. The settlement process has been underway for decades, so far paying out some NZ$2.272 billion in 90 separate settlements. Some settlements resulted in name changes. Aoraki, the highest mountain in New Zealand, was called "Mount Cook"—after the British sailor who landed in New Zealand in 1769—until a 1998 settlement with the Ngāi Tahu. (The official name is now the slightly unwieldy "Aoraki/Mount Cook.")

The Ngātiwai are the kaitiaki, or guardians, of kiore, MacDonald explained. The rodents are among the taonga that they should, by rights, have authority over. They want to see the kiore persist into the future. MacDonald doesn't buy that they are invasive because he doesn't really buy the idea that humans and all their works are unnatural. "We are less obsessed about that than Western conservation," he said. "Nature is just us." MacDonald added that if the animals were plentiful and nice and fat, my hosts might cook some up for me. MacDonald described them as "half the size of a New York sewer rat—all nice and fluffy and tasty looking."

Before we were allowed to embark, we had to submit to a biosecurity check by the DOC. A young woman unpacked and inspected all our gear in the back room of one of their offices, looking for introduced pests. Everyone was cordial, but there was a hint of tension in the air. After all, this ranger had the power to tell MacDonald he couldn't visit an island that his people claim as part of their territory and that they officially manage.

After an hour or so, our sailboat arrived at Mauitaha, which is steep and barely has a beach. It seemed to rise like a pillar from the ocean. There was no harbor to speak of, so we rowed to the shore in a small inflatable dingy through five-meter-high swells. Once we and our gear were on land, the captain of the sailboat waved goodbye and motored off with plans to return for us in a few days.

Waiting for us at a cozy campsite about 100 feet straight up were Mere Roberts, an anthropologist specializing in mātauranga pūtaiao (Māori scientific knowledge); Hori Parata, an environmental resource manager for the Ngatiwai; and two young "lads" learning the ins and outs of natural resource work, Te Kaurinui Parata (Hori's son) and Sonny Poai Pakeha Niha. One of the lads shouldered my pack and scampered up the cliff. I followed, considerably more slowly. When Roberts stopped me to point out a seabird burrow, I was exceedingly grateful for the excuse to rest. She pointed out a few feathers scattered outside the entrance to the burrow, indicating that it was occupied. Apparently kiore can co-exist with this bird—the oi, or North Island muttonbird—at least.

At the campsite at last, I was touched to see that the two apprentice natural resource managers had set up an area for my tent and demarcated it with a screen or curtain woven of flax. Hori Parata was sitting by the fire, his long hair pulled back in a ponytail and eyes hidden by a ball cap and oval spectacles. He was sipping one of the approximately 40 cups of tea he seems to drink a day. He told me a bit about Ruanui, the captain of one of the canoes to first reach Aotearoa, and the man who brought the kiore to the islands. They were an important food, he said, served to important people on important occasions. "A delicacy," MacDonald put in.

For years, the New Zealand Department of Conservation (DOC) had been eradicating rats from small islands all along the country's coast. Norway and ship rats had already displaced kiore from much of the mainland, so these islands were the last redoubts of the fluffy rodents brought over in voyaging canoes all those hundreds of years ago.

Roberts had been studying the kiore for decades. In the 1980s she roamed the Pacific, sleeping on the deck of a mail boat, eating raw tuna for "breakfast, lunch, and dinner," and trapping Pacific rats and preserving their livers in alcohol. Then she published a series of scientific papers using the data as a proxy for human migrations. Along the way, she grew fond of the rats. One day, Hori Parata walked into her office "in gum boots and overalls" and said, "I want you to help me save the kiore." Together, the two brought their concerns to the DOC, which in 2010—after twenty years of advocacy—promised to leave Mauitaha alone.

Parata remains aggrieved at DOC and at New Zealand's government. His management of kiore is thus both a way to enact ancient kaitiakitanga responsibilities and a way to protest against the management of his traditional lands by colonizers. "We are rangatira," he said. "We are not subservient and say, 'Please, Mr. White Man, can I go to my island?'" He wants there to be enough kiore for "cultural harvest." "We don't believe in putting them behind bulletproof glass," he said. "It is a food to us; it is a resource."

As the night fell, we heard the calls of Aotearoa's native owl, the ruru, which sounds exactly like its name, and which likely eats the kiore living on Mauitaha.

The next morning, we went to check the traps. The going was tough, because everything was straight up or straight down and the vegetation was thick. The views from the ridges, however, were worth it—vertiginous and emerald. From the top of the island, I could see Mauitaha's little brother, two-hectare Araara, which looked like the cover of a fantasy novel, a great castle of rock emerging from a foaming sea-glass-green ocean. I was feeling elated, but the mood darkened when trap after trap turned up empty. My hopes to eat kiore were dashed. Everyone was worried. Were the kiore all gone?

Months later, I received an email from MacDonald, assuring me that on a follow-up trip, the traps were full. Our timing just hadn't been right. Mauitaha isn't big, but it is big enough to keep a population of kiore alive, for now.

———

Ward and Campbell are both sure that what they are doing is not only right, but righteous. They have complete moral clarity. Ward sees killing exotic predators to clear the way for the resurgence of New Zealand's native birds as an unmitigated good, a wonderful activity that should be shared with the whole community. Campbell sees extinction as a moral evil that he is willing to dedicate his life to fight. "An extinction is bad—full stop," he told me. He calls the Predator Free 2050 "awesome" and expects that it may yield transformative innovations. Both of these men have weighed the price of saving species in the suffering of rats and other introduced species and found it acceptable. They sleep the sleep of the just. And, hanging out with them, it was hard not to get excited by the species that were coming back, the possibilities ahead.

But not everyone was as enthusiastic. Chris Thomas felt the quest to save hyper-specialized island animals was understandable but probably pointless in the long run. Wayne Linklater and Jamie Steer thought Predator Free 2050 represented an expensive overgeneralization about what kind of ecological management would best save New Zealand's iconic species. Kris MacDonald and Hori Parata experienced the quest to eradicate the kiore as colonial oppression.

It was time to get another perspective, to talk to scientists who thought killing for conservation was unjustifiable specifically because of the pain, suffering, and loss of autonomy inflicted on individual animals. They were waiting for me across the Tasman Sea on the world's biggest island, ecologically speaking: Australia.

13

Compassionate Conservation

A rian Wallach, a small, fierce Israeli ecologist working in Australia, is in many ways the anti-Campbell. Wallach believes killing in the name of conservation is never acceptable—and she says there are smart ways to protect both endangered species and individual animal lives. She's taken up the cause of defending non-native animal populations, such as camels in Australia and wild burros in the American West, populations that have been targeted for eradication. I went to Australia to meet her.

It was June and winter was coming on as I gazed out the window of a Saab 340 twin-engine turboprop, which seats around 30. I was one of seven passengers headed to Coober Pedy, an opal mining town 525 miles north of Adelaide. Below me, I saw apricot-colored earth with spots of crimson and coral and taupe, an archipelago of scrub in a sea of sand, overlaid by roads. The sandy soil in the Australian Outback is so old it has literally rusted. Then I began to see white cones, like small volcanoes, each with a hole alongside. These were mountains of spoil created by opal miners, just a few of the two million open mine shafts around the town.

I had a few days to spare in Coober Pedy, a parched town shimmering hot even in winter, laid out on two roads and featuring rusty spaceships from old movie shoots, broken-down busses, a drive-in movie theater, and plenty of opal stores, including a combination opal store/breakfast spot called "Waffles and Gems." The average summer temperature in Coober Pedy is in the high

90s, but the town was founded by miners, so many of the homes and business are underground, which keeps them cool. I checked into a subterranean Comfort Inn run over by an Evangelical Christian family and presided over by their curly-haired dog, who lounged around the lobby all day like a tiny potentate.

The dried-blood red color of the desert is apt. Australia rivals New Zealand in terms of killing animals. Though the desert landscape can seem lifeless from the plane window, there is in fact a rich and rapidly changing ecology here, featuring native animals, including kangaroos and emus, introduced animals such as camels, cats, foxes, and donkeys, and an animal that is in some liminal area in between—the dingo. And humans kill all of them—some to eat, some for conservation, and in the case of the dingo, to protect sheep.

Dingoes are very closely related to dogs. Whether they *are* dogs is a matter of some debate. Some experts use the scientific name *Canis familiarus*, implying that they are essentially free-roaming puppers. Others use the name *Canis dingo*, suggesting they are a distinct species. Those who favor *Canis dingo* say that dingoes are only halfway dogs, that they "split off before going through that full domestication pipeline," as one researcher put it in an interview with *Undark* magazine.

Human beings first discovered Australia 65,000 years ago. Dingoes came much later, likely brought by humans in boats from Southeast Asia some 3,500 years ago. (The arrival of the dingo is actually the only evidence that Australians had any contact with other people before the arrival of European explorers and Indonesian sea cucumber fishers about 400 years ago.)

Dingoes are compact and taupe, or black and tan, or occasionally white, with triangular ears and almond-shaped eyes. They are slim and long-legged. As they walk, they step in their own footprints, like cats. When European explorer William Dampier saw them in 1688, he described them as "little hungry wolves, lean like so many skeletons."

Although dingoes were brought to Australia by humans, today, they generally live on their own. But as dingo expert Bradley Smith writes, they "accept humans as social companions, but only if raised from a pup (around five to seven weeks) and provided with a suitable environment." In the old days, Aboriginal people used to pluck dingo pups from the wild. Some were eaten, but others were raised in the community, even breastfed by women. These young dingoes would act as "living blankets"—keeping humans warm on cold nights—and as hunting companions. Women, in particular, hunted with dingoes. Dingoes also helped humans find underground water, using their keener noses. But when dingoes reached sexual maturity, they typically left their human families and struck off on their own to find a mate. So they weren't pets, exactly, more like long-term guests, in a pattern very similar to the one reported for some Indigenous people and wolves.

European settlers started building fences to keep out dingoes and rabbits in the 1860s, and in the 1940s those fences began to be joined up into a single, 3,500-mile-long wire barrier designed to protect sheep in the country's southeast. The "dog fence" has been maintained ever since. I decided I needed to take a gander.

The desk clerk at the Comfort Inn recommended I call Wayne Borrett, who takes tourists out on jaunts into the desert. Borrett drove me north, out of town and into a red wilderness so vast and empty it made my molars ache. About half an hour later, we pulled off the road next to a slightly saggy mesh fence, maybe five feet high. We got out to have a look. Its posts looked to be tree branches, and Borrett confirmed that they are mostly fashioned from the local mulga trees, which are few and far between. Looking at the fence, I was surprised that it works to keep out dingoes. If they are anywhere near as good as escaping as wolves are, a puny fence like this should be no obstacle. I imagined they could easily leap over or dig under.

I said as much to Borrett, who laconically gestured at a sign affixed to the fence:

! WARNING !

POISON LAID

GREAT DANGER TO DOGS

RESTRAIN OR MUZZLE AT ALL TIMES

1080 MEAT BAITS PRESENT <u>AT ALL TIMES</u> TO CONTROL

WILD DOGS & OR FOXES ON THIS PROPERTY

TRAPS MAY ALSO BE SET

Oh. I finally got it. The Dog Fence isn't really a physical barrier. It is a target for the pilots who airdrop 1080 poison in a swathe of death on each side of the fence. It is the poison that stops the dingoes.

———

A couple of days later, Arian Wallach arrived with her team: Adam O'Neill, her partner and scientific collaborator; and two of her students, Erick Lundgren and Eamonn Wooster. Lundgren is an American who is studying feral donkeys; Wooster is an Australian with an image of his study species, the red fox, tattooed on his leg.

Our destination was Evelyn Downs ranch. Wallach comes to this massive 888-square-mile ranch because it is one of the few places in Australia where people aren't actively killing wild animals. Tough, outback Herefords share the landscape with kangaroos, wild horses, wild donkeys, camels, emus, cats, foxes, native rodents, dingoes, and very large prehistoric-looking reptiles called perenties. Of the animals on this list, dingoes, cats, foxes, horses, camels, and donkeys are all killed in large numbers throughout Australia—but not here. Wallach has convinced the owners to experiment with a more hands-off approach.

On our way into the ranch, we stopped at a water hole to check its level. Across the sunken puddle, a buff-colored donkey with an elegant neck stripe trotted down the slope to drink. Then a larger, darker donkey showed up and pulled rank, braying and snorting and claiming the right to drink for itself. The researchers watched, rapt.

Donkeys were imported into Australia in the late-19th and early-20th century from Spain, Mexico, India, Sumatra, and Mauritius as pack animals. Some escaped, some were lost, and some, their usefulness replaced by machines, were simply released into the desert. The latest government estimate, from 2011, puts the feral-donkey population of Australia at five million. Non-native and taking forage coveted for stock, they are considered less than worthless by conservationists and farmers. Agents employed by ranchers, national parks, and state governments frequently shoot them from helicopters.

Before leaving for Australia, I read a report from the Australian Broadcasting Corporation documenting that 6,000 wild buffalo, horses, donkeys, and pigs had been culled from Kakadu National Park in 24 days. I absorbed the information, but felt little. Sitting amid the flies and the heat, watching the soft muzzle of a caramel-colored Nubian wild ass ripple the surface of the water, I realized that I did not want this individual animal killed. But one donkey doesn't change the environment. Government ecologists contend that in their free-roaming tens of thousands, non-native herbivores graze and trample the desert to dust and turn watering holes like this one into mud puddles.

Wallach and her students say donkeys and other non-native species have been scapegoated. Lundgren says he has been unable to find any studies showing that feral donkeys endanger any other species. Their effects on landscapes are common to herbivores of any description, native or invasive. And the specific effects of donkeys are almost certainly swamped by those of cattle, which are "far more visible and pervasive, given their high stocking densities."

Wallach and Lundgren collaborated with Wayne Linklater and Jamie Steer from New Zealand and seven others to look at the species that have moved to Australia in the last several centuries. The team found that 23 of the newcomers are currently extinct or endangered in their native ranges, including the camel, a bovine from Southeast Asia called the banteng, and a little deer from the same region called the Indian hog deer. And yet most of them, like most introduced species, are considered "pests" and are not included in any measures of the "biodiversity" of Australia. They are seen simply as living pollution. Another way of looking at it, the paper's authors wrote, was that "the diversity of migrant species reflects a diversity of ecological threats and opportunities." Introduced herbivores can even perform the functions of herbivores that went extinct in the Pleistocene, like reducing fires by eating all the fuel or digging wells for water that can be used by other species.

In cases like the feral donkey, where most conservationists see killing non-native species as an obvious way to help restore a degraded landscape,

Wallach sees a puzzle to be solved. Step one: Stop overstocking cattle. Step two: Stop killing dingoes. If you leave them alone, they might prey on the donkeys and keep their numbers down. Do this and the ecosystem will sort itself out—no guns required.

———

The phrase "Compassionate Conservation" dates back to a 2010 symposium in Oxford, England, hosted by the Born Free Foundation—Elsa's legacy. In 2013, the Centre for Compassionate Conservation at the University of Technology at Sydney was founded. This is where Wallach works.

The philosophical underpinning of this approach is a work in progress, but Wallach is interested in virtue ethics. In a 2018 paper, Wallach and her colleagues wrote that "a virtuous person will carefully attend to the capacity of others to experience both joy and pain and make efforts not to inflict intentional and unwarranted suffering as a manifestation of one's compassionate character."

So far, the core work of compassionate conservationists has been to critique traditional conservation for being just as domineering and human-centric as the rapacious exploitation of natural resources it seeks to supplant. From the compassionate conservationist's perspective, both resource extraction and traditional conservation seek to control the non-human world and mold it to suit human values—no matter what the cost.

This critique has included identifying alternative narratives, win-win approaches where species are saved but humans don't do any killing. For example, where elephants in Kenya are being killed because they destroy farmers' fields, the compassionate conservationist promotes a fence that incorporates beehives, since elephants hate bees. (As a bonus, the farmers can collect honey.)

In another case, conservationists were killing foxes on Middle Island in Australia because they were eating rare little penguins. The colony, once 600 strong, had been reduced to fewer than 10 birds. But the island was so close to shore that even after all the foxes on the island had been killed, new foxes could simply walk over to replace them at low tide. A local chicken farmer who used a white, fluffy maremma sheepdog named Oddball to protect his flock suggested using guard dogs to protect the penguins. So in 2006, maremma sheepdogs were introduced and trained to guard the penguins. By 2017, the colony was back up to about 180 birds. The experiment even inspired a 2015 family film, *Oddball*. (Unfortunately, in August 2017, the penguins returned to the island earlier than expected and before the dogs were on-site. Foxes killed 142 penguins, setting the project back.)

Often, advocates say, a solution can be found by examining what all the species in the area want, what they are thinking, and how best to tweak their behavior. "It is actually a profound thing to realize that ecology is a bunch of sentient beings interacting in a landscape," says Wooster. "They are not just eating and fucking machines."

So far, this seems like a straightforwardly good idea. Why cause pain and suffering if there is another way? But I am interested in the hard cases where there is no other way, where conservationists really do have to choose between killing off thousands of animals and letting a native species go extinct. Then what do they do?

After our moment with the donkeys, it was time to set up camp. Wallach directed operations. She makes a lot of very direct eye contact, her green eyes blazing with intensity under the keffiyeh she wore in the field to keep off the salt-hungry flies that swarmed around us during the day. O'Neill cooked her the exact same breakfast every day in camp: veggie sausages, sautéed peppers, and a potato pancake.

Wallach was born in Haifa in 1977 and grew up in Jerusalem, New York, and Switzerland. After earning an undergraduate biology degree in Israel, where conservation focused mostly on preserving habitats for wildlife in a crowded country, Wallach chose to go to graduate school in Australia, a country she saw as wild and wide-open. In the mid-2000s, she began graduate research at Arid Recovery, a fenced reserve in South Australia. "It was the first time I encountered what I now call the dark side of conservation," Wallach says. "I thought conservation was a pure good."

Wallach was shocked and confused by the amount of killing that went on at the reserve, but found herself drawn to a man responsible for untold thousands of animal deaths: Adam O'Neill, who at the time was killing rabbits at Arid Recovery for a living. A commercial hunter and professional "conservation eradicator" with little formal education, he once invented and built a cat trap that used a trip wire to trigger a 12-gauge shotgun. But many years in the field had led him to a theory that so possessed him that he wrote a book about it in 2002. If humans simply stopped killing dingoes, he proposed, Australia's top predator could keep cat and fox numbers down, allowing native animals to thrive and humans to retire from shedding so much blood. By the time he met Wallach, he was seriously questioning the evidence behind the projects he worked on and was almost ready to quit as a gun for hire.

O'Neill is 16 years older than Wallach and speaks with a broad Australian drawl that contrasts with her carefully enunciated international English. He looks genuinely weather-beaten in a way she does not. He hunts kangaroos for his own consumption—a more ethical source of meat, in his opinion, than industrially-raised cows or pigs—and smokes an endless string of hand-rolled cigarettes, both habits that Wallach disapproves of but tolerates. At their home in rural Queensland, the couple share their lives with two rescue dogs and a large flock of free-range laying chickens.

After making camp, the team began setting up camera traps to record the behavior of dingoes, cattle, cats, foxes, donkeys, and camels around water holes. We spent a day setting out dozens of trays of peanut pieces, covering each tray with a layer of sand. Once the local rodents discovered the trays and got used to raiding the peanuts, some trays would be spiked with the scat of various animals to test whether native rodents were afraid of cats, foxes, or dingoes. The idea is that if the smell of the scat makes the rodents anxious, they'll spend less time at the tray, which can be measured by how many peanuts they eat before they scamper off.

With this project and others, Wallach and her team are probing how native and non-native animals interact, gathering data to make the case that they can and should coexist. Wallach believes that many native species can learn to live with non-native cats and foxes, but in Australia this is a minority position.

Like donkeys, horses, and camels, cats and foxes were introduced to Australia by settlers. These feral predators seem to eat just about anything bigger than three-quarters of a pound and smaller than 12 pounds—what Australian ecologists call the "critical weight range." Inside this range is an array of animals with wonderful names and uncertain prospects, among them the elegant predatory marsupial known as the spotted quoll; the bilby, which looks like a lean, erudite cousin of the rabbit; the chicken-like mallee-fowl; and the mini-wallaby known as the woylie. Every night, foxes and cats kill untold numbers of native animals and together they have helped drive more 25 species extinct, including the unsurpassably adorable big-eared hopping mouse. Thus governments, conservationists, and some landowners in Australia not only shoot large herbivores but also shoot, trap, and poison cats and foxes.

It is impossible to get accurate data on how many animals are killed in Australia. The federal government doesn't collect statistics on animals killed by its own departments, and the various state and territory governments don't collect data either. A spokesperson for the Queensland state government gave me a typical response: "Feral-animal control is scattered across so many agencies [that] it would be very difficult to quantify with any accuracy." One academic effort to get a handle on the scale of control determined that few records are kept, and that "institutional memory about pest control activities declined sharply after 5 years and was almost nonexistent after c. 10 years." Animals, in other words, are simply killed and forgotten and no one checks to see if their deaths accomplished anything.

The state of Victoria did provide me with one arresting statistic: Each year on average, they pay landowners and hunters almost a million dollars in bounties for fox scalps—at $10 per scalp.

In 2015, the Australian government pledged to kill 2 million feral cats by 2020. As of June 2018, they have killed an estimated 844,000 cats. (Their report on activities through June 2019 conspicuously didn't include a number of culled cats, suggesting the government may not be on track to hit their goal.) Camels are shot from the air. Rabbits are culled with poison and deliberately introduced illness—rabbit calicivirus. This is the kind of work O'Neill was doing when he first met Wallach.

Shortly after they met, Wallach and O'Neill started discussing ecological theory. And they started dating, making a compelling pair—a highly educated, cultured Israeli and a country boy with excellent aim. Katherine Moseby, the lead scientist at the Arid Recovery Reserve, remembers visiting the couple at their shared research campsite in a quiet corner of the desert. "We are sitting there after dinner just looking up at the stars. It was a beautiful night. And Arian said, 'And now I will play the harp.' And she went to her four-wheel drive and pulled out this harp—like this full-sized harp. And she sat down next to the fire and she played her harp."

In one of their first scientific collaborations, Wallach and O'Neill compared populations of species in different arid areas in Australia. On the allegedly dingo-free side of the Dog Fence, they found that thriving populations of yellow-footed rock-wallabies and malleefowl—animals vulnerable to cats and foxes—were living alongside dingoes. Stable dingo families— families whose breeding pairs weren't repeatedly disrupted by killing— seemed to be the best at suppressing cats and foxes. In 2010, a study with two colleagues found a similar pattern at more sites and with more native animals. Not everyone is convinced, but the results suggest that in some contexts, dingoes can protect some native species.

Wallach and O'Neill told me that after they published their results, government officials went to those sites and poisoned the dingoes. In their opinion, this is likely to doom those populations of yellow-footed rock-wallabies and malleefowl.

Even if they do convince their fellow Australian conservationists that dingoes can help protect some native species, it is by and large illegal to move dingoes to the notionally dingo-free side of the dog fence, and socially difficult to convince anyone to stop killing them anywhere. So this approach has never really been tried. Conservationists have continued to focus on creating fenced reserves free of all predators. The reserves produce generation after generation of bilbies, bettongs, wallabies, and other adorable creatures that are unprepared for life on the outside. When their populations grow too large for their fenced areas, they are sometimes "reintroduced into the wild," where they are promptly killed and eaten by cats and foxes.

If these reserve-raised creatures could be released into an area whose predators were regulated by stable dingo families, Wallach and O'Neill think they would be able to hang on long enough to adapt to the new ecology of Australia—to wise up and learn how to raise families out in the big bad world.

Other ecologists say that this optimism simply isn't supported by data. Moseby agrees that killing dingoes has "made it easier for cats and foxes to wreak havoc," but adds that even a single cat can take out a hair-raising number of native animals, so suppressing their numbers may not be enough. "I just don't think that it is as simple as adding dingoes and everything will be fine."

Wallach agrees that more research would help clarify the strength of the dingo effect. But despite the continent's wide-open landscapes, she says, it is nearly impossible to find a large area where dingoes have been left alone for long enough to settle down into a stable social-ecological relationship with cats and foxes. This is why she wants to found a Compassionate Conservation Research Station, a piece of land at least as big as Evelyn Downs, where the new ecology of Australia could be observed with minimal human intervention—and definitely no killing.

Wallach and O'Neill envision a future where dingoes, cats, foxes, and native species coexist, without human meddling, in a new ecosystem alongside other introduced animals like camels and donkeys—a radically different definition of success than that of the conservation mainstream, which instead seeks to preserve native animals in their native ranges, sticking whenever possible to the way the world was before humans changed it. In the ecosystem Wallach and O'Neill envision, cats and foxes would still be killed; they would just be killed by dingoes, not by us. This is fine with Wallach, because dingoes don't live by human ethics.

The landscape at Evelyn Downs is bone-dry but hyper-saturated with deep color: red earth, twisted khaki trees, a sky so blue it is almost cobalt. The ground is dotted with chunks of mica and, practically everywhere you look, stone tools. After all, people have been part of this ecosystem for over 50,000 years. Our group found many tracks and small scats but the only animals that showed themselves to us were the large ones who did not fear us—tough, placid cattle turning their white foreheads toward us, bounding kangaroos that raced the car down the road, horses running full speed across the desert on their own mysterious business.

Wallach set up an outdoor shower which pumped water heated over the fire up through a hose. A naked body dries almost instantly in this desert, so towels weren't necessary. One morning, we took turns washing up to pay a social call on the ranch next door, owned by the Lennons, an Aboriginal-Irish family that doesn't kill dingoes. They are "our companions," Bill Lennon explained, as we sat around his kitchen table. He wore a ball cap and a ring set with an opal the size of a gumdrop that was found on his land. "That's our spiritual side," he said. "Dingoes are a quiet animal," he said. "If you've got a dingo that has been with you for yonks, they'll only make this noise if there is danger." He produced a low, soft sound of canine concern deep in his throat. The only other time you'll hear them, he said, is when they are "answering each other" across the vast distances of the desert. "To me, it is music to my ears."

Lennon doesn't think all killing is wrong. He hunts kangaroos to eat on occasion. His description of the rules around hunting reminded me of other Indigenous hunting traditions half a world away from the kitchen where we sat and chatted. "We don't skin them and hang them up," he said. "It is a ritual, how you cook them. If we don't treat the animal the way it should be, the spirit won't provide another one."

After our chat, Lennon walked us out. Outside his house, dozing in the sun, was a tabby cat with a twisted leg, clearly broken and then healed at a 90-degree angle. Lennon explained that when he was out of town, a relative had allowed a government team to set out some traps for feral cats. This one got caught, but escaped. Lennon admits that cats can become "a bit of a nuisance," because of their appetite for native birds but it pained him to see the cat's broken leg. "I don't like killing animals," he said.

———

After a day of setting up experiments, we sat around the campfire, warming ourselves on a cold desert night as the Milky Way garlanded the vast dark above our heads. I challenged Wallach with the problem of Gough Island, a tiny speck in the South Atlantic which is home to almost the entire breeding population of the critically endangered and breathtakingly beautiful Tristan albatross. (The Tristan is very closely related to the more widespread wandering albatross and the birds are very difficult to distinguish visually.) The numbers of this huge, long-lived white bird are declining at three percent per year, in large part because their chicks are eaten alive by descendants of house mice that came to the island with people before 1888. Harrowing video footage shows the mice climbing all over the enormous fluffy chicks and literally eating them alive, nibbling into their skin and drinking their blood, while the parents look on, seemingly upset but unsure what to do. For millions of years, their kind has had no need to defend their chicks from any predators, and they simply don't know how. Population modeling suggests that unless something is done about the mice, the Tristan albatross will likely go extinct within a century. The Gough Island Restoration Programme, organized by the Royal Society for the Protection of Birds and a consortium of partners, intends to distribute brodifacoum poison across the island in 2021. If she could, would Wallach really cancel the operation? Would she be willing to see that entire species snuffed out so she would not have to kill any mice?

Wallach found my Gough Island question simplistic. First, these kinds of cases are uncommon, she said. Most species are declining for multiple reasons—including habitat loss and human actions—not just because of a single animal predator. And ultimately, what gives us the right to be the gods of Gough Island, to say who lives and who dies? The mice "aren't our children that we can control," Wallach said. "They aren't our pets or our livestock. They have their own agency. Conservation is ultimately a chauvinist method that treats animals as automatons."

So the Tristan albatross should be allowed to go extinct to preserve the agency of the birds and the mice? I find this hard to take.

Wallach sighed. She's seen the kind of death that poisons can deal out. Her own dog, Kuda, accidentally ate some 1080 bait and died screaming in her arms. Kuda's death was "by far the worst thing that ever happened to me in my life," she said. And she just can't imagine dealing out that kind of death—whether via 1080 or brodifacoum or some other means—to any creature. "I would never do that to a mouse. I don't care whose blood they are sucking on."

In the end, if it comes down to it, Wallach believes that individuals are more important than species. Individuals can feel pain and pleasure, can suffer and feel joy. Species are simply an abstract construct. Yes, the mice cause the chicks to suffer, but the mice are not bound by human morals, and we are not their masters. Because humans understand ethics, we have an obligation not to hurt or kill, but those rules only apply to us. It doesn't matter that humans brought mice to Gough Island. Those humans are long dead, as are those mice. The current generation of mice were born on Gough Island and have just as much right to be there as the albatrosses.

Although the world is filled with suffering, it is ultimately only human-caused suffering that Wallach seeks to eliminate—though she does still feel empathy for animals that suffer at the claws and teeth of other animals. "It is the very fact that life is composed of suffering that creates the necessity for compassion," she says. But she believes that when mice eat albatross chicks or dingoes kill cats, that is essentially not human business.

James Russell, the University of Auckland ecologist who has made a career of studying the control of introduced species calls this "a very naive view of human responsibility." He believes that we all have some obligation to undo mistakes made by our fellow humans in moving species around, and that there's little moral difference between killing an animal and letting an introduced predator do the killing. "We have to accept that [humans have] a larger collective responsibility," he says.

Russell isn't the only conservationist actively arguing against compassionate conservation. "The philosophy is not compassionate when it leaves invasive predators in the environment to cause harm" wrote 35 conservation biologists and ecologists in 2019. A 2020 paper with 36 authors predicted "dire consequences for global biodiversity" if compassionate conservation were to catch on. " 'Compassionate Conservation' may offer compassion to some individuals of a limited group of taxa, but ultimately consigns many more individuals to an uncompassionate demise."

The situation is somewhat analogous to the trolley problem. In some ways, using dingoes to kill cats and foxes is like pulling the switch rather than shoving the man. The result is still death for the cat or fox, but we haven't set the trap, fired the gun, or laid the poison ourselves. Russell is saying that even though one upsets us more emotionally, each death is morally equivalent, whether it comes from us laying poison or from the snap of a neck in the jaws of a dingo that we have ushered onto the landscape.

Linklater disagrees. "Eating another living thing to survive is a deeply ecological relationship. It is a universal law, like gravity. Life is death is life. Killing because we want to satisfy a cultural value judgement about nature seems shallow by comparison."

If I can't prove that "biodiversity" has objective final, then perhaps I need to be willing to say that I just love Tristan albatrosses enough to be willing to kill for them. That feels frightening to say, but I think it is true. Science can't justify the deaths of the mice objectively and neither can I prove that the persistence of a species is objectively worth more than currently living, sentient individuals. Perhaps the only legitimate argument here is simply that I—and many others like me—badly want albatrosses and other species to survive.

There are things we all might be willing to kill for. Most of us would kill sentient being to save our own lives, to save the lives of our nearest and dearest. We routinely kill sentient beings that eat our crops or stored food or try to move into our homes. Would you kill to save the Tristan albatross? The takahē? The Floreana mockingbird? The Floreana tortoise? Does it matter if it's genetically unique or not? Does it matter whether it'll be able to survive on its own after the killing or if it'll be conservation-reliant forever? Does it matter if it sings in a minor key? Does it matter if it's beautiful?

Then again, there might be a third way. What if we could change animals to save them?

14

Bilby Thunderdome

I don't like killing things, but when you have worked with threatened species and you've seen them be annihilated by cats and foxes, you can either sit back and watch or you can do something about it," Katherine Moseby says. "I have seen firsthand the way that one cat can destroy an entire population."

Moseby runs Arid Recovery, a nonprofit conservation organization managing 30,000 acres of red dunes 225 miles southeast of Evelyn Downs in the rusty interior of South Australia. This is where Wallach first learned of the "dark side" of conservation. The undulating land is dotted with tough, thorny scrub and divided into huge fenced enclosures stocked with Australian animals, most of which are on the verge of extinction elsewhere because of human-introduced cats and foxes.

Feral cats are not different in any biological way from house cats. Our pet cats are only semi-domesticated. They are still very genetically similar to wild cats. Researchers say the "modest" changes in their genomes show selection for "docility, as a result of becoming accustomed to humans for food rewards." In other words, we didn't domesticate cats; they domesticated us. It's no wonder, then, that they can easily slip between worlds, king of the couch as well as killer of the desert.

Native Australian animals have long been protected on offshore islands and artificial "islands" on the mainland, fenced reserves like Arid Recovery.

It isn't ideal, and everyone in Australia admits it. The woylies, numbats, quolls, and other spotted, striped, fluffy, and adorable native species are basically living in glorified captivity—in really big cages. But there's no realistic short-term prospect of fully eradicating the millions of cats and foxes currently living in the Outback.

Some conservationists have tried to train native animals to fear cats and foxes and run from them. They've tried squirting them with water guns while exposing them to a stuffed fox. They've shown them cats and then seized them and put them into bags, hoping to associate the predator with the traumatic experience of being captured. These efforts have sadly met with little success. Some never learn, and those that do fail to pass the lessons on to their offspring.

One look at "prey naiveté" around the world estimated it takes around 200 generations for prey to learn how to respond to a new predator. If the predator is capable of eating all the members of the species before appropriate anti-predator behavior evolves, you probably have an extinction on your hands.

At Arid Recovery, scientists are hoping to control this process carefully to avoid extinction. A small number of cats are being introduced into a fenced-off section of the reserve. The idea is to maintain their numbers at levels such that they will kill many, but not all, of the native animals there. The survivors will breed and reproduce. Most of them will be killed too, but, again, not all. Over the generations, only the animals most able to survive cats will live to breed, and eventually, the hope goes, a cat-savvy population of marsupial will emerge.

Arid Recovery operates on land donated by the multinational mining company BHP, which owns a neighboring uranium and copper mine. On my flight to the mining town of Roxby Downs were 38 men and six women. At the airport, my five fellow women and the herd of men in heavy work boots scattered directly from the tarmac to waiting trucks in various states of dustiness. I spotted my hosts: Dan Blumstein, an animal behavior expert from UCLA in glasses, a ball cap and lightweight scarf, and Mike Letnic, an Australian field biologist in shorts, rubber boots, and an extremely battered wide-brimmed felt hat. He smiled a wide, gap-toothed smile and welcomed me to Roxby Downs. My hair crackled in the dry heat.

The field station bunkhouse was full up, so Dan, Mike, and I were invited to crash in a modular workers' house previously occupied by an all-Aboriginal mining unit. We dumped our gear and hopped into a frankly filthy Toyota Hilux to head to the reserve.

At the reserve, I met Moseby and Katherine Tuft, the station's general manager. We all settled down to talk and eat and drink on the deck of the station's bunkhouse. As the evening drew in, and the sun's intensity mellowed, the wildlife came out. Zebra finches flocked in the tree beside the deck. Then the bettongs emerged, furry mini-wallabies, also called boodies, about the size of a kitten. They have small heads, big fat spherical rumps, and round ears, like bears. And these bettongs, descendants of a group that lived on an offshore island, were complete strangers to fear. They smelled our dinner, and they moved in, unnervingly close. They hippity-hopped under the table, brushing against my boots and vacuuming up scraps of steak and bread.

Before the barbecue grill was even cold, they lined up to lick grease from its underside, their arched throats looking pale and very vulnerable. One even found Tuft's toddler's bottle and began contemplatively sucking down milk, its black eye meeting mine with mild interest.

It was surreal to be essentially mobbed by a protected species. It was as if a bunch of panda bears emerged from the trees at the park and started creeping toward you, eyeing your sandwich. Or imagine being hustled for your ice cream by a bunch of spotted owls with no respect for your personal space. Inside the fence, though, these creatures are a dime a dozen.

The sun was setting, so Blumstein, Letnic, Moseby, and I walked down a trail in the sand to a kind of observation platform elevated above the great rippling dunes of the seemingly endless desert. Button-like bunches of pale yellow grass and a few green trees broke up the red expanse. An enormous red sun lit up low clouds along the horizon a maraschino cherry red. A wind kicked up, blowing Moseby's long honey blonde hair behind her, and the temperature dropped about 15 degrees over a minute or two.

It looked like a wilderness but it was in many ways a zoo. The native animals, lovingly tended to, protected by fences, had no real idea that they were in Moseby and Tuft's care. And all the two women really want to do is put themselves out of a job.

The next day, I went out with the team to check on one of the cats being used in the evolution experiment. It had a radio collar, but it still took us hours to track it across the dunes.

Underneath a dead tree, Letnic found the remains of a wedge-tailed eagle's dinner, white teeth and an upper jaw, which he extracted from a gray ball of fur. They belonged to a bettong. It is so dry here that everything dropped simply sits on the sand, seemingly forever: from bettong bones, to dead wood, to stone tools from pre-European Australia, neatly knapped, still holding a sharp edge. While the red sand outside the fence mostly bears the prints of rabbits and cats, the dunes inside are inscribed with tracks in the indigenous language: the long heart-shaped back feet of the boodie, the sideways V of the bandicoot, the distinctive toenail marks of the greater bilby. The fences clearly make a difference here, but will these track makers ever be able to leave the preserve?

Finally we found the cat, holed up in a twisted snarl of dead and living scrub. Its pugnacious expression looked like that of any house cat, gazing at you with ancient hauteur from inside a cardboard box.

That night, I went out with Moseby, Letnic, and Blumstein in the Toyota Hilux to spotlight bettongs. Letnic pointed a bright handheld light out the window. In the 10-square-mile area with the cats, boodies scampered out of the way of the dusty pickup, their butts like furry bouncing balls. Letnic seemed worried that there were too many cats; the eyes of the feral felines shone in the spotlight, and the night seemed full of them. Some of the cats have nicknames given by the researchers. They are named after composers—Beethoven, Wagner. "We want them to create this . . . symphony of chaos," Moseby said. One agile tabby leaped over a saltbush, disappearing behind a dune. If too many cats reproduce in the enclosure, all the native species will be killed. If there aren't enough, the natives won't adapt. It is a delicate balance.

As we passed into the smaller cat-free zone, the boodies seemed notice-ably more dim-witted. Several times the truck was forced to stop while someone got out and tried to herd them out of our way. Letnic ran at a pair who gazed at him blandly. As he approached, they began running companionably along with him, the man and marsupials looking like three friends out for a jog. In the end, Letnic had to nudge them off the road with the side of his foot. Outside the fence, they would be cat snacks by now.

The difference between these naive animals and the marginally more wary bettongs in the enclosure next door probably still represents learning, this early in the study, but the team is also absolutely interested in using the cats as a kind of evolutionary filter. Smarter, faster, bigger, warier bettongs will survive the cats' wiles and predations and reproduce. Over the genera-tions, they should become able to coexist with cats.

"It might take 100 years," Moseby said.

Is this kind of manipulation of a wild animal acceptable when extinction looms? Moseby feels the crisis demands action and intervention, "not sitting back and saying, 'Let's let everything eat each other and see what is left.'"

One advantage of Moseby's approach is that if it does work, the animals will be able to look after themselves in the future and will no longer be "conservation reliant." Even a massive predator eradication on the scale of Predator Free 2050 in New Zealand implies that humans must continue to guard against new introductions of the predators. Here, if it works, the bettongs and bilbies should be set, even if humans stop caring about biodi-versity—or stop existing.

I've argued that "genetic integrity" is a false value, like other purist atti-tudes that seek to freeze life in one state. If beloved biodiversity is on the line and the real-world suffering of animals is minimized, the fact that the animals are changed shouldn't by itself make this route unethical. In a scenario where the bettongs fill up their giant open-air cage with babies and polish off all the food, who's to say dying by starvation would not be worse than dying by cat? But this project opens the doors to serious questions about changing the animals we seek to save. How far are we willing to go?

Moseby is working with simple tools—cats, fences, radio collars, and traps—but she's ultimately trying to change their genome. And she's tentatively interested in the genetic tools on the horizon. A "gene drive," if it works, could leapfrog 100 years of learning and evolution and death at the sharp end of a cat's teeth.

———

"Genetic engineering" of animals arguably began with artificial selection of domestic animals, since the practice of breeding for desired traits changes the genome. Genetic engineering of animals in the lab goes back to the 1970s.

But for decades, the practice has been too difficult and expensive for the chronically cash-strapped conservation field to even think about it. Then, in 2011, Karl Campbell was looking for better ways to deal with introduced predators on islands and stumbled upon an idea that smelled like the transformative innovation he had been looking for.

An entomologist at North Carolina State University named Fred Gould had written a paper positing that genetic engineering techniques that had been used with insects were ripe for deployment in other troublesome species, including rodents. (Along with driving island species extinct, rats and mice eat enough rice each year to feed 180 million people, and they transmit Lyme disease and hantavirus.) Scientists could use genetic engineering to favor certain traits, Gould pointed out, and push them through wild populations. Normally, for any given gene that comes in different types, an offspring has a 50 percent chance of inheriting the mother's version and a 50 percent chance of inheriting the father's version. But some genes have evolved a way to cheat this system—if one parent has the gene an offspring has a virtually 100 percent chance of inheriting that version. That mysterious cheat code is called a gene drive, and if scientists could engineer a synthetic gene drive, they could spread a desired trait through a population and down through generations. To eradicate rats on an island, you might push a gene for infertility that would cause a population to crash once it reached a certain prevalence—no poisons necessary. The rodents would simply fade away, like a dynasty with no heirs.

Campbell invited himself for a visit to Gould's lab in Raleigh. As you do, Gould turned to the Internet to figure out who Campbell was. "I was just shocked," Gould says. "If you look at the Island Conservation website it is all woodsy-greensy." A lot of passionate environmentalists are opposed to genetic engineering. Gould asked Campbell, "Do you know what you are getting into?"

Campbell did. He didn't care that other conservationists considered genetic engineering too risky to attempt and too unnatural to countenance. He wanted to stop extinctions. Gould liked the man's pragmatism.

Gould's ideas were theoretical. But in 2012 the prospect of making the theoretical real suddenly got a lot better with the discovery of the CRISPR technique, a new way to edit genes quickly, cheaply, and precisely. With CRISPR, any DNA sequence could be neatly cut and pasted into any location in any genome.

About two years later, Kevin Esvelt, a geneticist then at Harvard University, put gene drives and CRISPR together. Instead of poking a big fat glass needle loaded up with synthetic DNA into every organism that you want to change, you do it once, with a gene drive that encodes not only the gene you want but also instructions to do that same manipulation with the CRISPR technique in another genome. So when your altered organism mates, its chromosome gets to work, engineering the chromosome inherited from the mate too. This guarantees that the offspring has the desired change, plus the instructions to make the desired change. When the offspring reaches maturity and mates, the process repeats. In a perfect "global" gene drive, 100 percent of offspring have the gene drive carrying the desired trait.

The possibility was a tantalizing one for conservation. You could start thinking way bigger than Floreana: you could eradicate rats on the Galápagos island of Santa Cruz, with its 12,000 people. Or, hell, cats or foxes in Australia. You could fix every island in the world.

The idea of using gene drives to save species began to hum. Campbell helped organize people from Island Conservation and researchers in the United States, Australia, and New Zealand, as well as the United States

Department of Agriculture, to research the approach. The group formalized as the Genetic Biocontrol of Invasive Rodents program, or GBIRd. In June 2016, Paul Thomas, a mouse geneticist from the University of Adelaide, Australia, visited Gould in North Carolina and got fired up. Thomas felt that his lab could be the place to figure out how to make a synthetic gene drive work in rodents. If he could succeed in lab mice, he could succeed with the wild mice and rats that eat the eggs and young of rare species on islands. Thomas joined GBIRd.

When I visited Paul Thomas' lab in Adelaide, I accompanied a grad student named Chandran Pfitzner to the mouse rooms. Before entering, we put on blue suits, hair nets, and masks. Pfitzner sprayed down my notebook with antiseptic and led me down a warm, hushed hallway to a room full of Plexiglas mouse boxes on racks. The rooms were surprisingly quiet, almost muffled, with the merest undertone of animals burrowing and gnawing. The research mice smelled like sweet sawdust and salt. Pfitzner, consulting his notes on the cracked screen of his phone, plucked one up by the tail, grabbed a miniature hole punch, and awkwardly excised a tiny circle of skin out of its ear. The mouse didn't make a sound.

This mouse was created in another building on campus. There, a fertilized egg was pierced with a glass needle and injected with the necessary ingredients for overriding the random chance of inheritance: the molecular "scissors" used in CRISPR engineering, a guiding molecule that tells it where to cut, and a promoter to activate the scissors in the right tissues. In this case, the CRISPR-snipped gene was not for infertility but for coat color. The idea was to test the synthetic gene drive out on a trait that can be checked at a glance. If the drive was working, the mouse would be albino. Instead, it was a rather lovely taupe. Pfitzner put the mouse back in the box.

After we left the mouse room and stripped off our protective gear, Pfitzner popped the piece of ear skin under a microscope. He wanted to see if the elements of the gene drive were in place. The scientists had also inserted fluorescent proteins next to the "scissors" and other components, and the mouse flesh glowed with two colors, bright red and a neon green, under an inverted fluorescence microscope. All the pieces were there, but the taupe coat was proof that the elements weren't functioning.

Out of 30 mice, Thomas and Pfitzner did get three dark-gray mice with patches and sprays of white, suggesting that the drive worked in some, but not all, of their cells. "It is early days," Thomas said, gazing rather forlornly at a picture of a mosaic mouse that he printed out for me.

In 2019, a team at the University of California, San Diego, announced that they got a gene drive to work in a large percentage of female mice, though the males were unaffected. It may not be as easy as the geneticists originally hoped, but Thomas has no doubt his team, or another, will crack the code. It's simply a matter of time.

Thomas and some colleagues in applied math modeled how long it would take to eradicate an island mouse population of 50,000 by introducing just 100 mice engineered with a functioning infertility gene drive. The answer was less than five years.

In the ear-punched mouse, then, was the seed of an unprecedented possibility—that humans could not just change a few mice in an Australian lab but permanently alter all mice, everywhere. The 30-gram wriggler portends a kind of power over the world around us that we've never had before: an ability to edit—or to delete—whole species.

This potential means that Thomas is taking special precautions. He understands that it could be perilous to the environment—and would certainly be perilous for public relations—should a mouse with a drive toward albinism or infertility escape its Plexiglas box and start mating with the free mouse population. So the first thing he did was create a dedicated line of mice for these experiments. Thomas's gene drive will only activate in the presence of a unique chunk of bacterial DNA that was engineered into the hole-punched mouse and its companions. That way, if one of these mice slips out into the hills around Adelaide and mates with a wild house mouse, the gene drive won't kick in.

About five minutes after Kevin Esvelt invented CRISPR gene drives, he freaked out about them. The technology could do plenty of good by preventing the transmission of horrible diseases and controlling animal populations without any killing. But it could also—if used prematurely, greedily, or unilaterally—drive species extinct and destroy public trust in science.

Cerebral, willowy Esvelt is now a professor at MIT and looks as much like an indoor person as Campbell or Moseby looks like an outdoor one. When asked about the promise and peril of his intellectual creation, he brings up Boo, his rescue cat, who lost the tip of its ear to frostbite before being taken in. He envisions a future when a local gene drive could reduce feral cat populations, much in the way that Campbell wants to reduce rats on islands. "The thought of feral kittens freezing and starving to death is just viscerally painful for me," he says.

Note that he uses the term "local" gene drive. One of his responses to his freak-out was to come up with ways of containing synthetic gene drives to a set number of generations. He calls one approach a "daisy chain," which would add a sequence of genetic drivers that must be in place to propel the desired gene change. The first driver in the chain is inherited normally, not pushed into 100 percent of offspring, so when it dies out, the gene drive does too. Tweaking the number of drivers in the chain could theoretically allow you to match the size of the effect to the size of the population of creatures you want to get rid of on an island.

This daisy-chain method is still being tested in the lab, and Esvelt feels that, barring attempts to tackle global health crises like malaria, no one should try a gene drive in the wild until there is a proven local drive. Esvelt co-wrote an essay in which he responded to New Zealand's interest in using gene drives to eliminate introduced predators. He called the basic version of a gene drive unsuitable for conservation purposes and warned against its cavalier deployment. "Do we want a world in which countries and organizations routinely and unilaterally alter shared ecosystems regardless of the consequences to others?" he wrote.

Esvelt has the same qualms about GBIRd's early and enthusiastic interest in exploring gene drive technology. GBIRd has said that its members intend to pursue a "precision drive" approach, in which the drive would work only on animals with a specific genetic sequence—similar to the fail-safe system Thomas is currently using in the lab. Researchers would have to locate a

DNA sequence found only on the target island and nowhere else, a prospect Esvelt thinks is unlikely. "There is a high chance it won't work out and they are building up hope," he says. On larger islands, there would be too many genes coming and going from other places for a perfect sequence.

Although Esvelt supports species conservation, he believes ethical priority must be given to preventing human and animal suffering. "The risk is that you could potentially cause a tragedy in the form of an accidental spread that would delay the introduction of a gene drive to stop malaria," Esvelt says. "Sorry, I don't care about endangered species that much." He's more interested in stopping disease transmission and controlling populations of feral cats.

But he says he wants GBIRd to carry on—as openly and carefully as possible, and in consultation with the public—because he does care about the suffering of the invasive animals that are currently killed with brodifacoum and 1080.

Esvelt himself is working on a project to disrupt the cycle of Lyme disease in Nantucket, Massachusetts. The people on the island objected to using a gene drive, so the current plan Esvelt helped develop would simply swamp the local Lyme-susceptible mice with up to 100,000 mice engineered to be Lyme and tick resistant. The hope is that the resistance genes will spread far enough in the population to make a difference. He is willing to let the community set the pace.

Campbell insists that he and GBIRd are committed to being careful and deliberate. Pretty much voicing Esvelt's exact fear, he says, "If you screw it up the first time around, you might put it back 30 years." In the meantime, he waits and keeps poisoning things, hoping to stave off extinctions and make islands safe for species that remain.

Gene drives and other CRISPR tools could become powerful ways to shape the world around us. Beyond just making introduced rodents, cats, and other predators infertile, we might eventually be able to use these techniques to make species we want to protect better able to cope with the changing world. Imagine making the bilby and the bettong smarter overnight—or even just *bigger*—harder for cats to take down. Imagine making species stressed by climate change heat resistant. Imagine using a gene drive to remove horns from all rhinos so there would be no reason to poach them—and then using another gene drive 100 years later to put the horn back once the market has dried up. Imagine using a gene drive on ourselves to make us more ethical or more wise to help us make these hard decisions.

Now, more than ever, it seems prudent to critically examine our values and goals before we unleash these new tools. And the tools themselves may take on a life of their own, according to philosopher Ronald Sandler. "Technologies are not merely used in human activities to accomplish goals, they structure the activities," he writes. "They are not merely an efficient power to enable people to accomplish their ends, they reorganize social and political power. They do not merely allow people to realize their visions for how the world should be; they alter what their visions are. In these and other ways, technologies often reshape us, even while we use them."

A study of 8,199 New Zealanders found that just 32 percent support using gene drives to achieve the goals of Predator Free 2050—but overall the most popular response was "don't know." I, too, don't know.

At one level, it seems rather precious to be squeamish about meddling in the genomes of wild animals. The vast majority of animals on Earth, by weight, are cattle, pigs, chickens, and other species that we have substantially altered through conventional breeding. Was domestication of animals ethically wrong? If not, how could it be wrong to engineer a coral to be able to live in warmer water or to engineer a bilby to be able to outsmart a cat?

On the other hand, the virtue of humility is important. So we must carefully pause. Are we changing other species for us or for them? Are we preserving possibilities into the future or closing off possibilities for future generations? Is this a situation where the species we are worried about will be able to adapt to new conditions on their own, coming up with their own solutions, or are extinctions clearly going to occur without intervention? What does everybody who cares about this species and this place think about changing it to save it?

One possible consequence of changing species is that we may—accidentally or intentionally—change their "capacities" or "capabilities" as well. This is the conundrum explored in the subgenre of science fiction known as "uplift" stories—narratives in which animals have been given new abilities to manipulate objects, to think, reason, and communicate. The first and most famous uplift story was H. G. Wells's novel *The Island of Dr. Moreau*, in which a mad vivisectionist creates humanoid versions of animals who are both horrifying and pitiable, and the narrator must work out what he owes to them, if anything. "Before, they had been beasts, their instincts fitly adapted to their surroundings, and happy as living things may be. Now they stumbled in the shackles of humanity, lived in a fear that never died . . ." (He settles for freeing them from their sadistic creator, some mercy killing, and then fleeing the island and leaving the rest to sort themselves out.)

We have actually already done this to some domestic animals, as any dog owner knows. Different breeds of dogs have different innate desires—some to play, some to hunt, some to retrieve, some to herd, some to guard. Our dogs clearly want to do these things. Indeed, they may desperately beg to play catch or be simply incapable of controlling their urge to bark at strangers. And yet humanity has bred these goals and interests into them. We have, in a sense, created new obligations for ourselves by creating new desires in our pets.

If we alter animals in such a way that we change what it means for them to flourish, do we have a special duty to provide for that need that we have created? In the sense that we would become partial authors of sentient beings, it stands to reason that we would assume new, quasi-parental responsibilities. The analogy is precisely with domesticated animals, which we have so thoroughly changed that they are now dependent on us for their happiness. The comparison suggests a possible guideline for engineering wild animals: ensuring that whatever changes are made to save their kind from extinction will allow them to flourish independently of any human management, given that perpetual human caretaking cannot necessarily be counted on.

And yet, the opposite course—to engineer non-humans to flourish more easily, to be simpler creatures with fewer wants or, perhaps, a higher tolerance for pain—seems almost more monstrous, a lobotomization of an entire species just to spare our feelings of guilt or empathy. The repugnance many feel when contemplating either "uplift" or what we might term "downlowering" are warning signs that we are treading in very deep waters.

If we were to make bilbies and bettongs smarter to save them, would we then owe them an answer if they used their new cognitive power to gaze across the red dunes and wonder what, in the end, was the meaning of all our animal lives?

———

As in the northern white rhino project, conservationists are increasingly looking at the possibility of using genetic techniques to bring back extinct animals, from the very recently departed to those that were hunted by our Pleistocene ancestors, whether it is the Tasmanian tiger, the passenger pigeon, or the mammoth.

Most of us agree that we should not drag living things into the world merely for a lark. There must be a good reason for it, especially since experience with cloning suggests that there may be considerable attendant suffering—of the females of the related species whose eggs are harvested for the procedure, of the surrogate mothers bearing mismatched offspring, and of the imperfectly formed animals that may result from early attempts.

So, is there any ethical reason to resurrect lost species like the mammoth? One reason might be that the act of de-extinction is reversing a wrong, undoing a bad act that has occurred in the past. After all, most species that have been proposed for de-extinction were first ushered into oblivion with the help of humans. Thus the act could be viewed as an act of restitution.

But for de-extinction to function as restitution, it has to restore (or at least try to compensate for) what was lost to an injured party. And who is the injured party?

The intuitive answer is "the mammoth." After all, they were alive, roaming the tundra and feeling the sun on their coats, and now they are all dead. The problem is that it doesn't seem like one can have actual moral obligations to what doesn't exist. One might claim that the obligation is to the future members of the species. And for many environmental ethicists, obligations can absolutely be grounded in our duty to future generations of humans and other species that we know will exist. However, if humans decide *not* to revive the species, then they can ensure no morally relevant individuals will ever exist and thus ensure that no obligations will exist. If you don't want responsibilities to passenger pigeons, then don't resurrect passenger pigeons.

Another possible ethical grounding of de-extinction is that we owe restitution to the ecosystems that lost the extinct species. But as we've seen, we can't really "owe" an ecosystem anything. It isn't really a coherent thing, and it isn't sentient. It can't really be the object of moral obligations.

Perhaps we could instead owe restitution to the individual sentient creatures that are part of that ecosystem. This would include, for example, the other animals who lived in the mammoth's ecosystem and who now live diminished lives.

But even though the few thousand years since we lost the mammoth are the tiniest of instants compared to the age of the Earth, life is remarkably nimble, as we have seen. Species that were initially hit hard by the mammoth extinction have settled into new grooves or died out. Some are likely extinct themselves, such as mammoth gut and skin bacteria, parasites, and any commensal organisms. (Did the great beasts support any now-gone species of birds that flew after them foraging in their woolly hair and dung?) Other species adapted to changing conditions. Life moved on. Given these facts, it is hard to make the case that sentient individuals living today are suffering because they are forced to live in a world without mammoths.

Perhaps the wronged party is not the mammoth or its ecosystem-mates. Perhaps it is us.

We have been harmed by this extinction. Scientists have been denied the ability to study mammoth ecology and behavior. Many generations have lived and died without knowing the pleasure of seeing the distant sight of a single file of huge woolly mammoths, their tusks curved almost in a circle. And no one alive today has tasted mammoth steak.

But if it is just for us, is it really worth it? How many elephants will have to be artificially inseminated with woolly mammoth zygotes? I cannot shake the image of Chai the zoo elephant chained up so zoo staff could work a plastic tube of sperm into her reproductive tract. Can we really justify doing such things to living animals just because it would be cool to have mammoths around?

Perhaps the alternative is to commit to make enough space for non-human animals on Earth to thrive and to protect the planet itself for millennia to come, creating the conditions in which new huge fantastic beasts might evolve. If we took care of the habitat of our elephants and allowed them the freedom to roam and explore, might they not one day, many generations hence, find their own way north again, gradually adapting to the chill, growing woolier, expanding out over a tundra refrozen firm by the concerted will of a planet of caring, humble humans who have worked to reverse climate change?

———

I once met a baboon researcher at a faculty party after I gave a lecture at a university in the Midwest. I asked about her baboons, and she described their society, which is strictly hierarchical, male dominated, and routinely violent. When low-ranking males and females step out of line, a giant male baboon will bite and beat them until they submit. Sometimes, baboons are beaten for no reason, just a periodic, random reminder of who is in charge. The victims generally take it, and then wait for their chance to usurp the leader and take over themselves. Males who want to reproduce will kill a female's infants or beat her until she miscarries to speed up her sexual receptivity. Females have their own pecking order too, and low-ranking females are the last to eat and drink. These primates are—let's face it—ass-holes. And they are miserable. Lower-ranking males and females are constantly on alert, expecting the next assault. High ranking males are paranoid, expecting up-and-comers who want their position and access to mates. Females are chronically abused. As she described this primate dystopia, I had a sudden unprompted and rather shocking thought: Maybe they should all be put out of their misery. *Maybe this species just shouldn't exist.*

In a now-famous study, researchers observed one olive baboon troop that, for a while, escaped the cycle of abuse and misery. In the early 1980s, two baboon troops lived near a tourist lodge. The Garbage Dump Troop ate trash and defended the dump as their territory. The nearby Forest Troop mostly stuck to the woods but coveted the goodies at the dump. Eventually, the most aggressive males from the Forest Troop began making forays to the dump. They'd decided it was worth brawling with the Garbage Dump Troop to get the food there.

In 1983, these baboons ate some dump meat that was infected with tuberculosis. It killed off nearly all of the Garbage Dump Troop and the aggressive males from the Forest Troop that had taken to eating there. After the abrupt disappearance of the most aggressive half of the troop's males, the vibe changed in the Forest Troop. A more peaceful culture with a "relaxed dominance structure" and less random violence emerged. Females remained the majority and were gentle and welcoming to incoming males from other troops. Low-ranking males in the new regime were not balls of stress, which suggests that they were also healthier.

The new culture lasted for 20 years, long after all the original less aggressive males had died. New males migrating into the troop apparently took a look around and decided to alter their behavior to fit the norms of their new troop.

If we truly care about baboons flourishing, it seems possible that humans might have a duty to periodically go into every baboon troop on the planet and shoot the most aggressive males—or use genetic engineering to change their natures, "morally enhancing" them to be more loving and kind.

Why does this idea both appeal to me and horrify me? Lori Gruen tells a story in her book about Peter Singer. He was giving a lecture to college students when he mentioned that although intervening to stop predation would cause more harm than good, if he could stop chimps from fighting with each other, he would. Gruen, who counts Singer as a friend, was appalled. "[F]ighting among chimpanzees is centrally important to their social interactions. It allows them to establish or reestablish their social hierarchy and, once the fighting is over, to engage in reconciliation, which strengthens social bonds. Stopping chimpanzees from ordinary fighting and altercations (which the possible exception of fatal conflicts) would actually be contrary to their interests."

Here, Gruen seems to be supporting fighting on the grounds that it is natural. Reading this description of the role of fighting in chimps made me think about my own childhood with my two brothers. We physically fought often, in part to establish our social hierarchy. I was the oldest, and when my brothers didn't do as I said, I sometimes clobbered them. They would fight back, and they often initiated fights as well. My youngest brother, Alex, was absolutely fearless and without mercy despite his size. He'd go for your eyes, your hair, and he'd frequently use his head as a battering ram with brutal effect. Later we would reconcile and maybe that did strengthen our social bonds. We are good friends now and enjoy spending time together. Does that mean our mother should have just let us fight as much as we wanted? Does the fact that it came naturally to us mean that our mother splitting us up and sending us to our rooms was "contrary to our interests"? (My current relationship with my brothers is very important to me and I believe it enhances all of our flourishing. If you could prove to me that our childhood brawls were *necessary* for our current close friendship, I might agree that the price was worth it. No permanent damage was done.)

The naturalistic fallacy is mistaking that which *is* for that which *ought* to be. Martha C. Nussbaum isn't about to make that mistake. "The conception of flourishing is thoroughly evaluative and ethical," she writes. "It holds that the frustration of certain tendencies is not only compatible with flourishing, but actually required by it."

Determining what "flourishing" looks like for a non-human animal of a particular species should start with studying its "characteristic ways of life" now, but this picture of its life should be considered "the beginning, not the end, of evaluation," Nussbaum says. Nussbaum thinks animal autonomy is important and cautions against "benevolent despotism" because it is "morally repugnant." But there are limits to letting nature take its course. Nussbaum rejects the idea that we should not invene in the lives of wild animals simply because they are wild. She points out our "pervasive involvement" with all the Earth's ecosystems—the same realization that prompted me to reexamine my own laissez-faire intuition.

So what kinds of interventions are called for? Should we step in to stop predators from killing prey? For a predator, the suffering associated with being prevented from hunting and killing may well be considerable. But the gazelle in the tiger's jaws presumably does not suffer less because its death helps the tiger express its core capabilities. Nussbaum admits that "the question must remain a very difficult one." Ultimately, she does not have a satisfying answer.

And what if the individuals in a certain species are routinely cruel to other individuals of that same species, as baboons are? I get the feeling that Nussbaum would be as annoyed at those high-ranking bully baboons as I am. At any rate, she does not rule out intervening. "Animal cultures are full of humiliation of the weak by the strong and of sometimes violent competition for sexual advantage," she writes. "Probably this is a case in which we must say that only the most egregious harms to weaker species members must be prevented." More broadly, since the non-human world is not by its nature just, Nussbaum supports "the gradual supplanting of the natural by the just."

Nussbaum isn't the first person to imagine remaking "nature" into a kinder, gentler, place. When the eighth century BCE prophet Isaiah foretold the coming of the Messiah, he famously characterized his kingdom, Zion, as a place where the violence of the wild world is no more: "The wolf also shall dwell with the lamb, and the leopard shall lie down with the kid; and the calf and the young lion and the fatling together; and the little child shall lead them. And the cow and the bear shall feed; their young ones shall lie down together; and the lion shall eat straw like the ox."

Most have seen this as allegorical, but some, it seems, would like to make it reality. Today, "effective altruism" is a popular movement with a utilitarian worldview that asks individuals to focus on using their time and money to do the most good possible. Unlike Singer himself, many of effective altruism's younger followers think humans can avoid doing "more harm than good" when it comes to intervening in wild animals' lives.

Deputy Director of the Wild Animal Initiative Cameron Meyer Shorb says that only focusing on the harms that humans have done to other species isn't enough. Fixing those threats to animals' welfare makes us feel good about ourselves, but a wild kangaroo doesn't care if it dies in a fire caused by climate change or a "natural" fire. Both are equally horrific. A life in the wild, Shorb says, is filled with pain and suffering—fear of predators, injuries and illness, coldness and hunger.

The Wild Animal Initiative focuses on researching the welfare of animals with very large populations on the theory that improving the welfare of a million mice is better than improving the welfare of three tigers—logic almost diametrically opposite to the standard conservation point of view. One of their initiatives, for example, tries to convince cities to control their feral pigeon populations with birth control-laced birdseed rather than poison.

Oscar Horta, a philosopher at the University of Santiago de Compostela, writes that "most sentient animals who come into existence in nature die shortly thereafter, mostly in painful ways (due to starvation, predation, and other reasons). Those who survive often suffer greatly due to natural causes." Many fish species have thousands or even millions of offspring, only a few of which will survive to adulthood. "Many just starve or are devoured soon after they start to be sentient. As a consequence of this, their lives have few or no positive experiences at all. They consist only, or mainly, of the suffering these animals experience when they agonize and die." Horta calls this state of affairs "Natural Hell" and asks, "Suppose we could modify 'Natural Hell' in a way that could, if not eliminate, at least reduce the harms suffered by those in it." Jeff McMahan, a philosopher at Rutgers, concludes that "the natural world" is a "vast, unceasing slaughter" and argues that once our scientific understanding of ecosystems improves, we should consider peacefully bringing about "the extinction of all carnivorous species."

Thinking about this level of intervention into the wild moves us into deeply speculative territory, where our intuitions seem to break down.

Imagine a year 2322, when the problem of wild animal suffering has been "solved." All female animals are allowed exactly two offspring and then are painlessly sterilized. Baboons have been engineered to be kind, living in loving matriarchies and never suffering the loss of a child. Predators experience the joy of predation via virtual reality, sparing any actual prey from a grim death in the wild. Toads dine on soy-paste worms; tigers gorge on "Beyond Gazelle" cutlets. Animals wear monitors that alert a central AI when their stress hormone levels spike, and a team of robot helpers arrives to assist. When they grow old, drones euthanize them in their sleep. We are as gods, creating a safe and pleasant world for all species. Ecosystems as we know them have been dismantled.

Or perhaps predation and death have been allowed, but suffering has been eliminated. All animals have painkillers and tranquilizers implanted in pellets under their skin. When a wolf or lion seizes them, sensors determine that their race is run and the drugs are deployed. The carnivores still get to hunt and eat, but the gazelle goes out in a painless haze of morphine.

Such futures seem both compassionate and completely absurd—even faintly disgusting. The very idea offends our sense that humans should be humble and not seek to be masters of the whole Earth. That this vision is repellent to many should tell us something about what we would give up were we to try to control the biosphere in the name of benevolence. Ending ecosystems to end suffering is a trade-off I am not ready to make at this time in history. As a species, we are too ignorant, too likely to be in error. And I still feel like something of ineffable value would be lost—but what ineffable value lies in suffering and death?

The notion of sovereignty that we considered when weighing the pros and cons of feeding polar bears helps here. Nussbaum invokes it when arguing that although we have duties not to thwart animals as they seek to express their capacities, we don't have a duty to be "benevolent despots of the world." She is moved by the argument that "the sovereignty of species, like the sovereignty of nations, has moral weight," she writes. "Part of what it is to flourish, for a creature, is to settle certain very important matters on its own, without human intervention, even of a benevolent sort."

Minimal changes to animals undertaken to keep an evolutionary lineage going in the face of challenges caused by humans seems ethically acceptable, even good. If we could save the 'akikiki and other Hawaiian birds by using gene drives to stop mosquitoes from transmitting diseases to them, and we knew it would be safe and effective, and local stakeholders—especially Native Hawaiians—were in favor of the project, I'd support it. If we could safely use gene drives to cause rodent, cat, or fox populations to gently dwindle away to nothing on islands where they are causing extinctions— without any risk to mainland species—I would support that too. I'd also vote for engineering coral to withstand global warming. And I'd even support changing bettongs and bilbies if that was the best way to keep them around. Giving them a fighting chance without having to drench Australia in blood sounds like a win-win to me.

One way to work through the ethics here is to think about how the bettongs would vote on this question if they had cognitive capacities like ours. Let's say you are going to be reincarnated as a bettong in the next life. Knowing this, might you want the species to be engineered to be able to live alongside cats?

We humans are in the process of considering a similar question for our own species. The International Commission on the Clinical Use of Human Germline Genome Editing released a report in September 2020 in which they recommended that humanity extend the current moratorium on making heritable changes to our own genes—for now. (Scientist He Jiankui was jailed in China for creating the first gene-edited babies, in 2018.) But in the future, the report said, serious diseases like cystic fibrosis, thalassemia, sickle cell anemia, and Tay-Sachs disease could be eradicated using CRISPR. Whether to pursue that goal is something that must be decided by "individual countries following informed societal debate of both ethical and scientific considerations," the report's authors said. Some 72 percent of Americans think that this kind of use would be ethically acceptable. And a healthy 60 percent approve of using CRISPR to "reduce a baby's risk of developing a serious disease or condition over their lifetime"—a goal not too different, in theory, from reducing a baby bettong's risk of being eaten by a cat over their lifetime.

CRISPR is powerful, but it isn't quite ready to be rolled out as a conservation tool. Fortunately, there is still time for discussion and debate over its use. The only way these tools can be deployed in an ethical way is to make sure there is a completely open, transparent, and radically inclusive process to weigh the pros and cons and seek something like consensus. The interests of non-humans could be represented in such discussions by human proxies. If we manage to get the governance and decision-making process right, it'll be an accomplishment arguably more profound than the creation of the genetic technology it would regulate. The real transformative innovation would be taking power from the rich and powerful, who typically manage the non-human world for maximum profit, and learning to manage it for mutual benefit, all together, with care and love.

Back at the Arid Recovery Reserve, Moseby bends over and picks up something feather-light. She holds it out toward me. It is the mummified remains of one of her bettongs. Its metal ear tag is still attached, reading "3139." The animal is curled up in fetal position, its eyes closed, its skin dried and split in places. I can see its perfect white skull peek through. This desiccated corpse, weirdly, is the goal of all Moseby's work: a bettong that wasn't eaten by a cat, that lived its life and then died of something else. Native or introduced, predator or prey, genetically modified or not, there's one commonality to all our animal lives. They are temporary.

15

How to Be a Good Human to the Non-Human World

Once, I dreamed I was in a moonlit garden with a square pond bordered by lilies and rushes. From between two of these plants emerged a black snake with off-white crescents behind each eye. "You see," I said to someone standing near, "humans and animals can live together." The snake slid noiselessly into the pond and swam away.

The dream stuck with me the next day, and I wondered if somewhere in the world there existed a real snake that looked like the one in my dream. I searched around online for a while, then turned to Twitter—a virtual world where I confess I spend probably too much time. I wrote, "I had a really vivid dream featuring a beautiful snake: black head with a cream-colored patch behind its head on each side. Is there a real species like this?"

A bit later, I got a response from a climate activist in Denmark. "Yes! Natrix natrix. Congrats. Such a good friend to humans. Some people even allowed it to overwinter in their houses with them." A Czech zoologist wrote at almost the same time with the same suggestion: "Grass snake, Natrix natrix. Perhaps the most common snake in continental Europe."

I searched for an image of the species, and when the pictures popped up, I felt all the hairs on my arms rise. It was undoubtedly the snake I had seen in my dream. Grass snakes even hunt in and around the water. I have no memory of knowing about this snake before, although it is certainly possible. I have been to Europe several times and spent time in the countryside where this snake might be encountered. I may have seen it in a nature documentary long ago. It is possible that the dream surfaced a memory I have no conscious access to.

I wanted to learn more about the snake and, in particular, about its relationship with humans. I found a paper by two Dutch scholars, Rob Lenders and Ingo Janssen, which argued that the reason *Natrix natrix* is so common in Europe is that snakes of this species like to lay their eggs in warm cow pies and other livestock manure. They thus followed Neolithic humans and their domesticated animals up the continent, into colder climes than any other egg-laying snake, and became culturally associated with livestock, even thought of as their protectors. They were "considered to be chthonic deities"—gods of the underworld—"not to be harmed," the researchers wrote. Every spring, they emerged from their underground hibernacula before the snow had melted and were thus seen as "heralds of spring." Their unblinking eyes, the researchers wrote, "made them to our forebears all-seeing creatures and therefore very wise." The snakes also symbolized death and, as they shed their skins, rebirth. In the Baltic countries, grass snakes were sometimes considered to be the spirits of dead ancestors, "taking care of their descendants by protecting valuable cattle and by stimulating fertility." And Lithuanians and Latvians did indeed let grass snakes live in their houses, near the hearth, to keep warm, and sometimes fed them. Their worship goes back to Indo-European times.

The researchers speculate that the ring around the neck of the grass snake, which goes almost, but not quite, all the way around, may have inspired the common European Iron Age accessory known as the torc (or torque), a metal neck ring with a gap in the front. They point to the fact that many torcs were decorated with snake motifs, and they mention the Gundestrup cauldron—a huge silver bowl from sometime between 150–1 BCE found in a Danish peat bog. It is covered with figures and scenes wrought in silver, including the antlered god Cernunnos. He wears a torc and holds another in his right hand. In his left, he holds a grass snake. Around him are animals: deer, something canine, something feline, and a small person riding on the back of a dolphin.

I gazed at pictures of the cauldron, feeling like I was looking at scenes from a European dreamtime, when animal people and human people spoke to one another, when animal gods and animals themselves were worshiped, when humans knew themselves to be simply one kind of animal among many. A pre-Christian world before the human/nature duality.

Christianization more or less disrupted the worship of grass snakes, and they became despised—like all snakes—as symbols of the evil serpent that tempted Eve in the Garden of Eden. But in some corners of Europe, respect for this species persists. A Romanian naturalist writes that in Eastern Romania, "No one kills this snake because it is considered as a protector of the house, a help against mice and insects." In the Netherlands, volunteers now construct "broeihopen" for their local Natrix species to nest in: piles of loose compost that hold the heat to keep the eggs warm. We did not tame these animals and we do not routinely keep them captive, and yet our lives have been entangled with theirs for thousands of years.

And the grass snake is hardly the only example of such entanglement. The Yao tribe in Mozambique have a special relationship with a wild bird called a honeyguide. As its name suggests, the birds lead the humans to honey. The arrangement is mutually beneficial. The birds can find the hives, which are usually in hollow trees. When they find one, they rustle up a human and communicate with them, using a specific "chattering call." Once they have a human's attention, they lead them carefully to the hive, flying from tree to tree. There, the humans expose the hive and subdue the bees with fire. After the hive is on the ground, the humans take the honey and the birds feast on the wax. If a human wants to go honey hunting, he calls the birds in with a specific noise that scientists describe as a "a loud trill followed by a grunt: "'brrrr-hm'" Apparently, the Hadza people in Tanzania use a different noise, a "melodious whistle," to call the same birds.

In southern Brazil, wild bottlenose dolphins and fishermen work together nearly every day to catch mullet. The dolphins herd the fish toward the shore, where the fishermen await. Dolphins signal to the fishermen by sticking their heads out of the water and hitting the surface of the sea with their throats, in what the fishermen describe as a "head knock." At this, the fishermen throw their nets. The dolphins eat panicking fish fleeing backwards from the nets. Everyone wins. (Except the mullet.)

This isn't even the only place where humans and dolphins work together. Fisherfolk and dolphins also collaborate in the Ayeyarwady River of Myanmar. Here the humans call to dolphins by tapping their canoes or making "guttural sounds with their mouths," according to a report on the practice. Dolphins call to the humans by leaping from the water and slapping their flukes near their canoes, then lead the humans to schools of fish, which they then obligingly herd into the fisher's nets. Like their Brazilian cousins, the dolphins eat "fish that are either stunned or darted away from the sinking cast-nets," and also nibble on fish protruding from the nets.

These are unexpectedly good relationships, the kind I wish we had with more wild animals. Some of our closest cross-species relationships changed us both so much that our animal partners are no longer considered "wild." Hunting with wolves, sharing a home with wildcats—how else did dogs and cats become parts of our families?

––––––

We need to do a better job sharing with non-human animals. Just a quarter of the Earth's ice-free land is considered to have "very low" human impact, and nearly all of that is in the tundra, the boreal forest, or the hottest deserts. By weight, there are ten times as much humanity as wild mammals in the world. And our livestock weigh almost twice as much as we do. Seventy-five percent of all birds, by weight, are chickens. Thirty-eight percent of the Earth's ice-free land is covered in crops or pasture. The photosynthesis carried out by all those plants, which captures the Sun's energy, serves us. The sunlight that once fuelled millions of species increasingly pours down human throats, and we top it off by guzzling millennia of sunrises and sunsets in the form of fossil fuels.

With wild animals squeezed in to narrow slivers of space in between farms and roads and cities, conflicts multiply, populations become fragmented and endangered, and our ethical conundrums multiply. By simply reducing the human footprint and creating *more space* for other species, we can let them sort out many conflicts and solve many problems on their own rather than having to intensively manage them. Respect for sovereignty can become more common than compassionate intervention. Simply put, the more room animals have, the less micromanaging we will have to do and the easier our ethical decision-making will be.

We can make more room, and in some places we are already doing it. Some areas of the world are seeing an increase in wildlife as they transition away from widespread agriculture and toward a landscape in which much land is returning to a feral, unmanaged (and yet still deeply humanized) state. The whole U.S. East Coast has transitioned twice since European settlement. First, it went from a complex environment shaped by Native Americans that supported large numbers of forest species like deer and passenger pigeons to a cleared, agricultural landscape with few trees, eventually losing its iconic chestnut trees altogether to an introduced blight. But then farming moved west and a new, chestnut-free forest returned, this one a patchwork of trees interlaced with suburban developments and roads. With the return of the trees came the return of deer and bears and raccoons and coyotes (but no passenger pigeons—yet). Much the same has happened in Northern Europe, as agriculture has moved east and south.

Many of these trends are driven by outsourcing land-intensive agriculture to other regions. To shrink our footprint on a global scale, humanity cannot simply shift agriculture and infrastructure around; we must learn to shrink it. Technology will help here, but so will cultural change, including eating less meat, especially beef, since cattle take up a lot more room per calorie than plant-based foods.

Ultimately, we already know how to take up less space. What is lacking is the will. We can learn a lot from how Indigenous human societies on islands managed their environment before the global economy tied these places into mainland supply chains. As Sabra Kauka, a cultural practitioner who teaches Hawaiian Studies in Kaua'i told me, "An island is a microcosm or a magnifying glass." In pre-colonial Hawai'i, less than 15 percent of the land was transformed for habitation and intensive food production to fully support hundreds of thousands of people. Flood-irrigated terraced fields produced taro, fish, and waterfowl. Large chunks of forest were kept sacred. "Hawaiians are not above the land; the land is above us," she told me. Regulations for interacting with the land were detailed and if you broke the rules "the penalties were stiff." The results was 'āina momona—a fat land of abundance and diversity.

We can and should retain and expand large areas of undeveloped space where big animals can thrive. But that may not be enough. We also need to create ways to share space—to allow some species to geographically overlap with us—and accommodate them in our space. This is a key to being good neighbors on Earth, to reducing the number of conservation-reliant species, to respecting the sovereignty of non-human communities. We simply need to take up less space.

We must also meaningfully tackle climate change. When rising temperatures squeeze the ranges where plants and animals can survive, that is another way we take up space on planet Earth. We've talked about climate change mostly in terms of how it humanizes the world, increasing our ethical responsibility to our fellow species. But you should also fight climate change for your many fellow humans who will be harmed by it. It's a clear-cut case of injustice. The people who profited from burning fossil fuels can always fly further north in their private planes. The Earth's poor will be the ones who lose their crops, get sick, and die young from asthma. For all these reasons, any person worried about living an ethical life should figure out an effective and sustainable way to fight for climate justice.

This may honestly be the most straightforward piece of advice I can give you: Make room for other species and fight for climate justice. Doing just those two things will do so much good in the world. And they will help not just people and other animals but trillions of other living things as well.

———

There's a weird thing about loving the world, a thing that slithers under the surface of my thoughts as I hike through the woods or paddle down a river. Ecosystems are built on death.

The Amazon Rainforest is a place of great beauty, mind-boggling diversity, and adorable squirrel monkeys romping in fig trees taller than a ten-story building—but it is also a place where harpy eagles snatch baby monkeys off their mothers' backs, a place where the bullet ant produces pain so intense that being stung by one feels like being shot, a place where a jaguar will eat your chickens, your dogs, even your children, if you aren't careful.

In 2014, I interviewed John Terborgh at Cocha Cashu, the station where he had been conducting research for 30 years, a place he chose for his work because it had the "proper complement of animals"—it hadn't been hunted or logged or otherwise disturbed by humans so heavily that some species were gone. In his research on how ecosystems were structured, he found that predation was the most important organizing force. "Predators are just everywhere," he said in his office. He caught a cicada and held it as he spoke, between his thumb and forefinger, while it buzzed furiously. "Ninety-five percent of nesting attempts fail," he told me. In my memory of the interview, he says to me, "*All the babies die.*" But I can't find that in my notebook. Maybe that is how I summarized it to myself later, sitting under a fig tree, watching a dung beetle roll a wad of howler monkey poop into a perfect orb.

"Nature" is the interwoven web of life on Earth, a pattern made up of an uncountably massive number of individuals, each one the product of billions of years of evolution. And evolution carves its masterpieces out of death. Our species only exists—you only exist—because of random genetic mutations and the deaths of untold billions of individuals who either did not live to reproduce and contribute their genes to subsequent generations or *did* live to contribute their genes—and then died anyway. We stand on a pile of corpses going back 4 billion years to the Last Universal Common Ancestor. And in so many of those lives there was suffering—fear, grief, pain—oceans of blood.

When someone very close to me died, I remember walking around a lake, in disbelief. Sixty years of memories, an absolutely unique personality, tastes, opinions, regrets—a thing of such complexity took decades to create. It didn't seem possible it could disappear in an instant. The way he used to line up his spare change on the kitchen counter. Gone. If a person's self was knit together by memories and the enacting of a self over time, then it stood to reason that *unknitting* would take time too. But no. Death is easy, like slipping into the bath.

There are nearly 8 billion people on Earth. I cannot even understand that number, not really, and the vastness of it astonishes me every day. I told my kids once that 250 babies are born every minute on Earth, four every second. To this day, they sometimes stop in the middle of whatever we are doing—cooking dinner, walking to the corner store—and say "a baby is being born *right now.*" But the numbers of humans are nothing compared to the number of sentient beings—of selves—on Earth right this moment.

To hold in our minds both the inestimable value of each self and the massive numbers of selves that there were, are, and will be exceeds the ability of the human mind. We are used to common things being cheap and rare things being valuable. But selfhood is both common and priceless.

It is because death seems to vaporize selves, to unravel in an instant a lifetime of knitting, that we hate and fear it. Fear of death is the central theme of *The Epic of Gilgamesh*. After Enkidu dies—in part as punishment for killing Humbaba—Gilgamesh mourns him for six days and seven nights, refusing to let his corpse be taken for burial "until a maggot dropped from his nostril." After his death, Gilgamesh becomes overcome with fear, saying, "I shall die, and shall I not then be as Enkidu? Sorrow has entered my heart! I am afraid of death, so I wander the wild to find Uta-napishti." Uta-napishti the Distant, you see, knows where to find a plant that will make those who eat it immortal. After many adventures, Gilgamesh finds the plant, but while he is bathing in a pool of water, a snake steals it and eats it, sloughing off its skin and living anew, reborn. Gilgamesh weeps with anguish and goes home.

But death creates life too, as the maggot that drops from Enkidu's nose suggests. The dead are food for others. The soil from which plants grow is enriched by dead animals and plants, rotted into life-giving earth. And we all must die, contributing our bodies to the compost. As biologist Bernd

Heinrich says, "But in fact we do not come from dust, nor do we return to dust. We come from life, and we are the conduit into other life."

Philosopher (and crocodile attack survivor) Val Plumwood says that the two dueling modern Western ideas about death are both unsatisfying. Either our immortal souls go on to "continuity, even eternity, in the realm of the spirit" or, if we aren't religious, we see it as "the complete ending of the story of the material, embodied self." Believing our souls fly up to heaven like transparent cartoon angels leads to "alienation from the earth," while believing that we just die and go poof means "the loss of meaning and narrative continuity for self." She offers as an alternative Indigenous animist concepts of self and death that see life "as recycling" back to the "land that nurtures life." In Plumwood's metaphor, life is like a book "borrowed from the earth community circulating library." It might be marvelous, but it isn't yours to keep.

When we die, our consciousness goes out like a candle being blown out, but Plumwood questions whether consciousness is the only part of us that counts as our selves. We are also our bodies, which can be food for so many—for the condor, the wolf, the raven, the wedge-tailed eagle, the polar bear, the centipede, the lion, the fungus. Just after finishing her paper on death, but before it was published, Plumwood was found dead on her undeveloped property outside Canberra, and "it was suspected she had been the victim of a snake or spider bite," according to the *Sydney Morning Herald*. It would have been a fitting end, but it was actually a stroke. At least she was outside.

The interconnected network of life that the environmentalist in me worships was created accidentally by evolution, a totally amoral process which just happened to produce a world so beautiful that it can make you feel like your heart is exploding inside your chest. This world is constructed and nourished by the flow of matter and energy from the atmosphere and the sun into one organism and then into another, transferred by death and eating and dying and rotting. That *flow* itself feels like a valuable thing. In "The Land Ethic," Aldo Leopold defines "land" as "a fountain of energy flowing through a circuit of soils, plants, and animals." Death keeps the energy flowing, so it is necessary and valuable to the core work of conservation: keeping the flow going, keeping the relationships alive, keeping evolution going. Take the Gough Island case, the one I presented to Arian Wallach around the fire in Australia. Poisoning the island's mice can be seen, in this framing, to be the work of rerouting the flow so that it goes back into the albatrosses, which can then connect the land with the sea, extending the flow further. The agony and deaths along the way are sad—but all of us must die, all of us become food someday. Death and pain are not horrible aberrations; they are normal.

Caring for individuals is the opposite. The core work of caring is directing energy and matter toward one body, keeping it alive. When I care for my children, I feed them energy from other bodies, other lives. Three times a day, I route sunlight that has passed through dozens of species into their little mouths and it gives me a deep, ancient satisfaction to do so. Caretaking is about pooling energy and resources in beloved individuals, not keeping the flow going.

At some fundamental physical level, prioritizing *the flow* and prioritizing *the individual* cannot be reconciled. There is no way to care for all the individuals on Earth at once, since feeding one and keeping it alive means keeping it out of the stomach of another and thus starving the other. If we appoint ourselves the apportioners of all matter and energy on Earth, we take on a burden far too colossal to understand, let alone bear. We are neither worthy nor able to be managers of the whole Earth at once.

This is the tension I have been struggling to reconcile all this time—between my respect for the value of the individual and my more mysterious sense of awe at the flow that connects us as animals to all other living things—an awe that is central to my love for the non-human world. Plumwood grappled with this tension twenty years ago, writing, "Any society that values individual life forms highly will experience conflict with the radical flows of the food chain narrative." The only way to resolve these two stories, she said, is to "see the world and ourselves from both sides at once."

———

I am approaching the end of my journey, and I am beginning to feel like I've been keeping two lists: a list of things that are really *good* and a list of things that are not. Here are my lists. Let's start with things that are often claimed to be valuable in the environment that I don't think actually are valuable.

<u>Things that are not valuable</u>
1. Naturalness
2. Wildness (defined as lack of human influence)
3. Ecological integrity
4. Genetic integrity
5. Purity in general

All of these alleged values flow from human/nature dualism, which is false and damaging. The first two, naturalness and wildness, are different ways of saying "places and organisms untainted with human influence," which assumes humans are both unnatural and by definition destroyers of nature—neither of which is true. The value of "ecological integrity," like the related concepts of "native" versus "invasive" species, attempts to enshrine one state of a landscape as the morally correct or superior state. Since this state is nearly always defined as the pre-colonization or pre-human state, it again boils down to devaluing states or organisms tainted with human (or European human) influence. "Genetic integrity" does the same thing at the level of an organism or a species—defines the pre-human state as morally superior and deems any changes that flow from human influence as pollution.

Purity itself I add to the list to emphasize what a poor fit any type of purism is for ecosystems, species, or even individuals. By their natures, these things are dynamic, adaptive, interwoven and blurry-bordered. When people attempt to freeze the living world in time or put it in rigid boxes, it seems to me that they deeply misunderstand it. The protean nature of life is central to what makes it valuable. The living world we inhabit was created by millions of years of evolution—of change. In even a single generation of any given species, the genome changes. The only way to really stop life from changing it is to kill it.

Now, if you are arguing to preserve certain ecosystem states for *cultural* reasons, explicitly because a group of humans likes them or likes them a certain way, then that's different. Then you are simply saying that you want to prevent some change for the sake of human preferences. And if the costs are low in terms of suffering or harm, then I see no reason not to proceed. I manage my own garden based on my own preferences, killing and removing plants I don't like, adding and moving around plants I do, largely to please myself and my family, though with some thought to the needs and desires of the birds and insects with whom I share the space. If you manage a park and you are committed to a certain "natural range of variability," and achieving that outcome won't hurt anyone, then go nuts! Just don't argue that your goal is objectively valuable or somehow derivable from science. I think such gardening or landscape management is good and even necessary, as long as the whole Earth isn't managed. We must remember the virtue of humility. Ideally, on a large landscape scale, there should be both unmanaged places and places managed in different ways with different goals in mind. Such diversity of approaches allows us to compare outcomes, hedge our bets, and will allow us to honor the multiple ways that the non-human world can be valuable. And now my list of things that *are* valuable in the non-human world, in descending order of how confident I feel about placing them on the list.

Things that are valuable
1. The flourishing of—at least—sentient creatures, including their autonomy
2. Human compassion
3. Human humility
4. The flow of matter and energy between living things
5. The diversity of living things

Remember that flourishing is a broad concept; it goes beyond being "happy" to encompass leading a good, full life. What I am saying here is that individual sentient creatures matter. I could argue for this claim in a number of ways. I could argue, like an Aristotelian, that flourishing is just good. I could talk like a utilitarian about maximizing the preference satisfaction of morally considerable entities. I could talk about the rights of humans and non-humans to express their capabilities. I could talk about distributive justice—international, interspecies, and intergenerational—and use that framing to call for sharing the goods of the Earth with all humans and all sentient creatures, including those yet unborn. I think that any number of approaches will make the case. In many ways, this is the easiest value to justify. There is near unanimity that human flourishing, that human lives have objective final value—and since we are closely related to and cognitively and emotionally similar to animals, most of the arguments for human value also apply to animals.

Whether these same arguments apply to members of other kingdoms, notably plants and fungus, whose different ways of communicating and making choices in the world are only beginning to be understood, I leave for another day. But from evidence that some individual plants are more willing to take risks than others, to new research on forest-wide multi-species tree-and fungal networks, our non-animal kin are almost certainly both selves and networks in ways that will challenge and complicate our ethical intuitions. I look forward to digging into these questions in the future.

For many kinds of individual animals, I think autonomy is part of their flourishing. Being able to make their own choices about where to go, what to do, with whom to mate, and so on is part of what makes their lives good from their own perspective. We have evidence for this preference in the efforts made by Thelma and Louise to remain with the herding dog that had become part of their family, in the attempts of primates, elephants, and other species to escape from zoos, even in the willingness of mice to free other mice from small cages. Freedom is independently valuable to many non-humans, just as it is to humans. Freedom is part of flourishing. And this kind of freedom is how I prefer to define the word "wild." A wild animal wakes up in the morning and it decides what to have for breakfast.

For the virtues of humility and compassion, my argument is rather backward; when people act in accordance with these virtues, mutual flourishing tends to ensue, and the complexity of non-human life is generally speaking less reduced. Things seem to go better when we act with humility and compassion. These virtues also feel good.

For the values of flow and of diversity, I can stand on very firm ground by arguing for the considerable subjective final value found in the non-human world. Millions of people all over the world *love* the non-human world over and above what it does for them directly. This love is found across cultures, ages, income levels, and personalities. The value of a sun-dappled meadow that you walk through, picking a flower to put in your hair; the value of the lifelong memory of the sound of a whale breathing from across the water; the value of the smell of the woods after rain—summed across all the non-humans and humans who delight in these things—is as vast and deep as the ocean.

I strongly feel that the flow of energy in and the diversity of the world of living things also has *objective* final value—that even in a universe with no sentient creatures to value them, ecologies of living things are still morally valuable. But, as you will recall from my agonies in chapter eight, I cannot prove this. I can only hand you the lupine leaf. I believe that humility also means admitting the limits of our knowledge and ethical understanding.

You will note that the valuable things on my list are not in the same currency, so to speak. The first resides in individuals, which are easy enough to count, although some people may want to weigh individuals based on various cognitive capacities or the number of years left to their lives or some other variable. The second and third are virtues, which you can't easily count or measure but you can usually recognize easily enough in action. Ecological complexity can be measured in various ways, as can biodiversity, but these measurements have different units. And so, we are faced with the incommensurable. How do you weigh 1,000 individual lives against a species? How do you weigh the value of humility against the value of an ecosystem?

Kant thought all morality boiled down to one single principle: Don't use people as mere means to an end. Singer can theoretically solve any problem with the single currency of preference satisfaction. But my list doesn't reduce down to a single principle, value, or virtue. That's okay, according to the philosophers. It turns out I am a "value pluralist."

Holding more than one thing as valuable means that sometimes you have to choose between them. How are we supposed to do this if they aren't in the same currency? Ethicists often say we are supposed to use what they refer to as "practical wisdom," something Aristotle came up with, which means something like "the judgement of a thoughtful, knowledgeable person with experience making ethical decisions." In many ways, we are back to moral particularism, in which we are enjoined to pay careful attention to all the features of a particular case, develop defensible reasons for acting in a certain way, and then act.

Taken together, I believe these values suggest that in a humanized world, we owe non-human animals respect and compassion, plenty of space, a climate that is not changing too quickly, and—in some cases—intervention to help them deal with environmental challenges caused by humanity. What we owe any individual animal depends on how that animal flourishes. In many cases, the best way to respect them will be to back off and allow them to make their own choices, even if what they choose means that "nature" looks strange and new to our eyes. In some cases, our reverence for the web and flow of life may suggest to us that we need to hurt or kill animals to protect an evolutionary lineage or ecological complexity, especially in the unique situation of introduced continental predators threatening island-adapted species. I believe that this price is worth paying in some cases, especially where deaths can be quick and suffering minimized. But we must not use killing as our default, fallback position. We must not take life lightly. We must try other approaches; we must ask ourselves hard questions about the genetic distinctiveness of the lineages we are protecting. We must seriously consider what our actions are committing us to in terms of assisting or guarding vulnerable species for millennia to come. We must also seriously consider the possibility—as in the case of the hybrid barred and spotted owls, or the black cherry adapting to defend itself against insects in its new range—that other species are actually solving environmental problems on their own and our lethal intervention is counterproductive.

If I were to try to summarize my conclusions in an Aldo Leopold-style guideline, it would go something like this: *A thing is right when it promotes the flourishing and autonomy of living things, their diversity, and the complexity of their interactions—but where we cannot promote all these things at the same time, we must make our choices with care and humility.* It's a bit longer than Leopold's version, but a pithy axiom does not always make for a bulletproof ethical system.

When you are dealing with dueling, incommensurable values, deciding to take an action that allows humans or other animals to enjoy full, flourishing lives sometimes means you might reduce biodiversity and complexity and vice versa. And since you can't compare what's at stake through a common currency, you can't just opt for whatever will cause the "least harm." In philosophical circles, this is known as a "moral dilemma." Sometimes, no matter what you choose, you may not be able to avoid doing wrong.

The wrong that you do when you choose what seems like the least bad of a set of bad options is called the "moral residue." Recently, Chelsea Batavia and Michael Paul Nelson, both at Oregon State University in Corvallis, and Arian Wallach wrote a paper on moral residue. They describe it as "the moral requirements that are left unfulfilled in morally dilemmatic situations." If you choose to poison the mice of Gough Island to save the albatross chicks, the moral residue is the pain and suffering and death of the mice. If you choose not to poison the mice, the moral residue is the pain and suffering and death of the chicks plus the loss of the species of Tristan albatross. Acknowledging the moral residue, we feel responsibility and, Wallach and her co-authors suggest, we feel grief.

So here's how I would approach some of the tough cases we've discussed where it is impossible to act in a way that preserves or promotes everything of value at the same time:

1. Determine what *you* think is really valuable in the case. Dig deep. Critically examine your own assumptions about what is valuable—and why.

2. List out all the options for solving your current dilemma and the likely consequences for each option. Don't forget to include "doing nothing" as an option but remember that "doing nothing" doesn't win by default,

since "nothing" usually means allowing pervasive human influences to play out—influences for which you may bear some part of our collective responsibility.

3. Guided by the values you have identified, choose the least morally wrong option, using your rationality as well as your emotional responses. Note: There may not be a "correct" answer.

4. Own your choice by admitting that you made the decision based on your own values and accepting any moral residue.

5. If necessary, grieve.

Of course, most moral dilemmas of consequence are not likely to be given to a single person to decide. This template for decision-making should ideally be undertaken collectively by all the interested parties, perhaps including non-humans in the form of appointed representatives. Depending on the group, you could end up with a list of core values even longer than mine. On the upside, working in a group means that when you make your decision and take your action (or intentional inaction) you will have someone with whom to grieve.

There is no single happy ending for life on Earth just as there is no simple formula for acting ethically in a humanized world. We must do the best we can with multiple incommensurable values, then live with the choices we have made, the species not saved, the pain we caused.

Make room for others; stop climate change; fight for justice; be compassionate; be humble; admit you don't know everything. Make homes for snakes. Sit quietly in the light of the last hour before dusk, the shadows of the junipers long and the colors bent blue. Listen to the swallows call as they swoop above you snatching midges from the air and know that we are not alone on Earth. All you can do is your best, every day—until that day comes when you lay your human burdens down and become, as an animal, part of the great feast.

ACKNOWLEDGMENTS

I drew the epigraph for this book—"All flourishing is mutual"—from Robin Wall Kimmerer's *Braiding Sweetgrass: Indigenous Wisdom, Scientific Knowledge, and the Teachings of Plants*. I believe her call to build good relationships with our fellow humans and non-humans (rather than either continuing destructive interactions or withdrawing from other species in shame) is as close as we are likely to get to finding the "meaning of life." The acknowledgments section of a book is traditionally the place where authors throw off the romantic fiction of the single independent author to expose the dense web of mutualisms that really creates a book. In that sense, it is like burying one's hands in the soil of the forest floor and feeling and smelling the network of intertwined roots and fungal mycorrhizae that constitute and support a "single" tree. I'm writing this section in March of 2021—the very last words I'll write for *Wild Souls* and in some ways the most important.

First, I acknowledge that this book was largely written on the traditional territory of the Klamath and Modoc peoples—today, with the Yahooskin Band of Northern Paiute Indians, the sovereign nation known as the Klamath Tribes. A central concern of the contemporary Klamath Tribes is saving the C'waam and Koptu, two endangered fish species that live in Upper Klamath Lake and its tributaries. Today, they are actively engaged in C'waam and Koptu research and conservation—holding up their end of their reciprocal relations with the species. At the annual C'waam ceremony, held in very early spring, with snow often still on the ground, a pair of fish is blessed by tribal elders and the released into the beautiful Sprague River. C'waam and Koptu once fed the tribe during the hungriest part of the year. Although they are now too rare to eat, they are still thanked every spring. A common Klamath saying is "naanok ?ans naat sat'waYa naat ciiwapk diceew'a," which means "We help each other; we will live well." All flourishing is mutual.

Second, I thank the editors who commissioned and edited stories I wrote about people and animals that became the seeds of this book: Adam Rogers and Vera Titunik at *Wired*, Ross Andersen and Michelle Nijhuis at the *Atlantic*, Rob Kunzig, Rachael Bale, and Brian Howard at *National Geographic*, Alex Heard at *Outside*, Lauren Morello, then at *Nature*, now at *Politico*, Hillary Rosner under contract at *Boom: A Journal of California*, Ursula K. Heise and Jon Christensen at UCLA, and Michelle Niemann, then at UCLA, now an independent writing consultant and editor.

I want to doubly thank Michelle Nijhuis, my good friend, for emotional support and many excellent campfire conversations about conservation. Thanks also to Jack and Sylvia. Rest in peace, Pika.

I'd like to thank the mink I saw in along the Flat Branch in downtown Columbia, Missouri while walking my daughter to daycare in 2010, who appeared during a moment of self-doubt as if to reassure me that my ideas about "nature" in the human world were worth pursuing.

Thanks to a young man who came to see my lecture at the University of Oklahoma in 2013 and who came up afterwards and told me he would have just let the condors die in peace—and thanks to Zev Trachtenberg for inviting me out to give that talk. Thanks to Chris Luecke at Utah State University for inviting me to talk about tradeoffs between animal rights and conservation way back in 2017, kicking off this book project. Thanks to Jamie Lorimer at Oxford for inviting me to give a talk in the fall of 2019, forcing me to organize my thoughts toward some actual conclusions.

Thanks to Fred Swanson, Michael Nelson, and the staff at the H.J. Andrews Experimental Forest for offering me the Andrews Forest Writing Residency. I had to leave early, because my grandmother was dying, but the days I spent in your forest were magical and brain-opening.

Thanks to Charles Mann for encouraging me to delve deeper into Descartes' view of animals, and for being consistently encouraging and supportive of my work over the years—and for the delicious preserves. Thanks to Wally Sykes for telling me to get a dog. I didn't get a dog, but I think I understand what you saw as missing from my analysis of human-canine relations. Thanks to Michael Soulé, a man I admired greatly, for sitting down for a beer with me to talk through our disagreements—which in the end were pretty minor in practice.

Thanks to Sabra Kauka and Maka'ala Ka'aumoana, Native community leaders and teachers, for sharing their perspective with me on conservation on Kaua'i. Thanks to Sheri S. Mann of the Hawai'i Department of Land & Natural Resources for introducing us.

Thanks to Katie Cassel for giving me her invaluable guide to the plants of Kaua'i and to Jim Cassel for the fantastic interview about pigs. I wasn't able to squeeze it into the book itself but it helped me a lot. Thanks to Cali Crampton for taking me and my husband to Bird Camp. Thanks to André Raine for taking us to see the petrel burrows and dealing with our slow pace and endless questions. I have never been so muddy in my life. Thanks to Brian Vastag and Beth Mazur for welcoming us to Kaua'i.

Thanks to Badger Run Wildlife Rehab in Klamath Falls for rescuing a bat from my house. Thanks to my brother-in-law Kevin Maier for taking me out to see the humpbacks bubble-feed and my sister-in-law Shayna Rohwer for holding and feeding my baby while I did. Thanks to all the wonderful philosophers at the International Society for Environmental Ethics for putting up with my layperson questions and for being so kind and welcoming when my husband and I brought our kids to the annual meeting. In particular, thanks to Av Hiller at Portland State University for pointing me to some fascinating work on far-future responses to wild animal suffering. Thanks to Edward N. Zalta and all the contributors to the Stanford Encyclopedia of Philosophy, a fantastic and free resource for understanding philosophical ideas. Go check it out. Thanks to Clare Palmer for agreeing to an interview and reading the chapter on your work. Thanks to Heath Holden for alerting me to the loiborkoram story.

Thanks to Glenn Shepard for agreeing to accompany me and translate on my trip to Manú. I would have made an absolute hash of the whole endeavor without you and I learned so much on our very long canoe rides up and down the Yomibato, Manú, and the Madre de Dios. Thanks to Nancy Santullo of Rainforest Flow for logistics, guidance, and for your ongoing work keeping clean drinking water flowing inside Manú, thus saving lives. Thanks to Francisco "Pancho" Peuño Sorecco, Javier Shumarapague, Alejo Machipango, Alex, and Nicanor for helping me get where I was going—and back. I'll never forget the time when we were headed back to Tayakome from Yomibato and the little 16-horsepower Briggs & Stratton lawn mower engine on the boat quit (again) after dark, and we were just drifting down the river under the stars. I wondered to myself why I wasn't worried and then I realized that I was surrounded by the world's foremost experts on surviving in the Amazon rainforest, and that you all would keep me safe. I really enjoyed our time together. Special thanks to Orlando Cornejo Ylla, for cooking on our trip to Manú and for delicately waiting for me to leave before you slaughtered chickens. I'll never forget your fried plantains and your dancing skills. Thanks also to Elias Machipango Shuverireni, Martin, Thalia, Maria, Ismael, Samuel, Paulina, Victoria, Celsor, Marco, Luz Marie, Yoina, Maylli, Florentino, Mauro Metaki, and all the many other Matsigenka who welcomed me with masato and were so generous with their time and knowledge while I was in Manú.

Thanks to Karl Campbell for touring me around Floreana and not laughing—very much—when I got a mosquito bite on my eye and spent a whole day of field reporting looking like Popeye. Thanks to Claudio Cruz for showing me around his farm. I still think often of the frigate birds flying parallel to the surface of your pond, tilting their beaks down to sip water at high speed—a truly beautiful sight.

Thanks to Kris MacDonald, Hori Parata, Mere Roberts, Te Kaurinui Parata, Sonny Poai Pakeha Niha, and the staff of the Ngātiwai Trust Board. I learned so much on our short trip to Mauitaha. Thanks especially for the wonderful gifts, which I still treasure. Thanks to Wayne Linklater for reading two chapters and offering me some excellent comments and for inviting me to speak at Victoria University in Wellington. Thanks to James Russell for reading parts of the manuscript and disagreeing with me often about introduced species but always being extremely nice about it. Thanks to Danielle Shanahan for the tour of Zealandia. Thanks to Paul Ward and Tim Park for the tour of Polhill.

Thanks to Brent Marris, my fourth cousin and proprietor of Marisco Vineyards in Marlborough, New Zealand, for your amazing hospitality. I learned so much from our time together—not just about wine, but also about pigs, dogs, land management, and my own family roots!

Thanks to Arian Wallach, Adam O'Neill, Erick Lundgren, and Eamonn Wooster for accommodating my visit to their field site in Australia and for answering my endless annoyingly hypothetical questions. Special thanks to Erick for driving me back to town—I forget how many hours round-trip that was for you, but I definitely owe you one. Thanks to Wayne Borrett for taking to see the dog fence and other sites around Coober Pedy and to Bill Lennon for the impromptu interview.

Thanks to Katherine Moseby and Katherine Tuft for putting me up, showing me around Arid Recovery, and sharing your work with me. And thanks to Daniel Blumstein and Mike Letnic for letting me tag along during your research group meeting. Thanks to Paul Thomas for inviting me into your lab and to Chandran Pfitzner for taking me to see the mice.

Thanks to Peggy Mason, Rolf Peterson, Steven Wise, Donny Martorello, Wendy Spencer, Pamela Maciel Cabañas, Luigi Boitani, Brenda Guernsey, Ronald Sandler, Rachel Nuwer, Gayle Burges, Lisa Wathne, Mike Clark, John Farnsworth, Oliver Ryder, Chris Thomas, Lotta Berg, Joris Cromsigt, Jamie Steer, Fred Gould, Kevin Esvelt, Eric Sanderson, Justin Marceau, John Terborgh, and Chelsea Batavia for their generous gifts of time and expertise.

Thanks to the sugar ants, *Tapinoma sessile*, for continually seeking entrance into the shed where I work and crawling all over me and my computer, forcing me to apply my ethical theories to a very real case study. Thanks to the mother and baby skunks who live under my back porch for being the quietest and most gentle of housemates and being a living example of human-wildlife "coexistence."

Thanks to Russ Morgan. I didn't put your story in this book, but I learned a ton about wolf management from you. Thanks to all the readers who subscribed to my content on wolves that I published at the now defunct Beacon. A lot of that work laid the foundation for this book. Thanks to my Twitter followers for providing me with a virtual community to share my ups and downs with while writing and editing. In particular, thanks to Jennie Kaae Ferrara and Vojtěch Kotecký for responding to my tweet about my snake dream and identifying *Natrix natrix* for me and to Theodore Davis for confirming that the kōkako's song is in a minor key.

Thanks to my mother, Kathrine Beck, and my brother, Alexander Marris, for reading portions of the manuscript and for many useful conversations. Thanks to Yasha Rohwer for philosophical assistance and moral support. I love you, Sweeto. Thanks to Adele and Nicolas Rohwer for many excellent dinnertime conversations about what we might owe wild animals.

Thanks to Sarah Chase for frequently watching my kids and bubbling with me. Thanks to Finley and Lucy Chase for being bubble friends. Thanks to the scientists, doctors, nurses, med techs, hospital staff, pharmacists, and many others who fought the pandemic while I hid in my backyard shed, writing this book and keeping my family close.

Thanks to Brooke Borel for helping me in my search for a fact-checker. Thanks to Emily Krieger for rigorous and often very funny fact-checking. (Any and all mistakes that remain are, of course, my bad alone.) Thanks to Alexandra Elbakyan for research assistance. Thanks to Mia Kwon for her absolutely gorgeous cover design and Ashley Polikoff for copy-editing. Thanks to Rosie Mahorter at Bloomsbury for her deft publicity work and Morgan Jones at Bloomsbury for helping me round up all the images and permissions.

Thanks to my agent, Abigail Koons of Park & Fine for consistently wonderful representation and advice. The first time we met, after we left a restaurant, you leaned over and delicately replaced a purse strap that had fallen off my shoulder—and it was then I knew you were the agent for me.

Thanks to Ben Hyman, my editor, for just generally being a genius and instrumental in making this book flourish. In particular, I am in debt to your ability with titles. Some of the best chapter titles are yours, and after much back and forth, you produced the book's subtitle, which is perfect.

Finally, thanks to the staff and volunteers at Rogue Climate, the Southern Oregon climate justice organization for which I also volunteer. The Rogue Climate office ironically burnt down in climate-change fueled wildfires in the summer of 2020, but the organization flourished afterwards, like a serotinous ponderosa pine cone cracking open in the heat and spreading its seeds of hope. Sometimes the link between securing financial incentives for low-income Oregonians to install heat pumps and respecting the autonomy of wolves or pikas might seem tenuous, but the fight for climate justice is also the fight to create conditions in which individual non-humans can thrive. And by lovingly but insistently demanding a world in which we really take care of each other, in which we all flourish together, you give me hope every day.

NOTES

CHAPTER 1: THE FLIGHT OF THE ʻAKIKIKI

1   introduced to Hawaiʻi in 1928: Linn et. al, "Red-crested Cardinal"

1   50 species of birds: My account of the evolutionary history of Hawaiʻi's bird species is derived from Ziegler, *Hawaiian Natural History, Ecology, and Evolution*, 251–264.

2   between 800 and 1,000 years ago: Hunt and Lipo, "The Last Great Migration," 208.

2   trapped birds with bait and sticky sap: Ziegler, *Hawaiian Natural History, Ecology, and Evolution*, 328.

2   came with a British ship in 1826: Van Dine, *Mosquitoes in Hawaii*, 7.

3   to teach the younger birds the melodies: Paxton et al., "Loss of Cultural Song Diversity and the Convergence of Songs in a Declining Hawaiian Forest Bird Community."

8   "only about 60 wolves in the state: Oregon Department of Fish and Wildlife, "Oregon Wolf Population."

8   "and the American west": Oregon Wild, "Wolves Come Home to Oregon."

8   more than 1,000 miles: Kane, "California Wolf Is Back in Oregon."

9   he refused to walk into any of the traps biologists set: Marris, "Why OR7 Is a Celebrity."

10   90 percent of the deaths were caused by humans: Hebblewhite, "Wolves Without Borders."

10   at least seven of them died: Oregon Department of Fish and Wildlife, *Oregon Wolf Conservation and Management 2019 Annual Report*.

10   Data on 155 deaths: Cubaynes, "Density-dependent Intraspecific Aggression Regulates Survival in Northern Yellowstone Wolves."

CHAPTER 2: OUR ANIMAL KIN

16   Thomas Nagel published a paper: Nagel, "What Is It Like to Be a Bat?"

17   13 million years ago: Nengo et al., "New Infant Cranium from the African Miocene Sheds Light on Ape Evolution."

17 66 million years ago: O'Leary et al., "The Placental Mammal Ancestor and the Post–K-Pg Radiation of Placentals."

17 550 million years ago: Evans et al., "Discovery of the Oldest Bilaterian from the Ediacaran of South Australia."

18 Cambridge Declaration on Consciousness; Low, "The Cambridge Declaration on Consciousness."

18 as animal behavior researcher Frans de Waal points out: De Waal, *Are We Smart Enough to Know How Smart Animals Are?*, 49–50.

18 the captive wrasses evidently understood: Kohda et al., "Cleaner Wrasse Pass the Mark Test."

18 Each has a unique call: De Waal, *Are We Smart Enough to Know How Smart Animals Are?*, 262.

18 Parrots have names too: Morell, *Animal Wise*, 97.

19 Bats also produce individually unique calls: Vernes and Wilkinson, "Behaviour, Biology and Evolution of Vocal Learning in Bats."

19 has even been demonstrated in spiders, cabbage white butterflies, crickets: Hedrick, "The Development of Animal Personality."

19 and bees: Walton and Toth, "Variation in Individual Worker Honey Bee Behavior Shows Hallmarks of Personality."

19 Frogs, toads, salamanders, and newts: Kelleher et al., "Animal Personality and Behavioral Syndromes in Amphibians."

19 10 juvenile Eastern garter snakes: Skinner and Miller, "Aggregation and Social Interaction in Garter Snakes."

19 "They can tell others apart.": Morell, "Snakes Have Friends Too."

19 Rats can be introverts or extroverts: Žampachová et al., "Consistent Individual Differences in Standard Exploration Tasks in the Black Rat."

20 A captive African gray parrot named Alex: Morell, *Animal Wise*, 80–85.

20 Scrub jays: Ibid, 55.

20 carrying the water in their own mouths: De Waal, *Are We Smart Enough to Know How Smart Animals Are?*, 68.

20 Capuchin monkeys will share: Ibid, 135.

20 Rhesus macaques subjected to a fiendish experiment: Masserman et al., "'Altruistic' Behavior in Rhesus Monkeys."

21 Dolphins will lift: De Waal, *Are We Smart Enough to Know How Smart Animals Are?*, 133.

21 Orcas were observed doing something similar: Marino et al., "The Harmful Effects of Captivity and Chronic Stress on the Well-being of Orcas (*Orcinus orca*)."

21 Kuni found an injured starling: Preston and De Waal, "Empathy: Its Ultimate and Proximate Bases," 19.

21   to reach and rescue another rat trapped in a small tube: Das et al., "Demonstration of Altruistic Behaviour in Rats."

21   over half of the time, they chose to share: Bartal et al., "Empathy and Pro-social Behavior in Rats."

21   If we give that rat an anti-anxiety drug: Bartal et al., "Anxiolytic Treatment Impairs Helping Behavior in Rats."

22   A camera trap video that went viral: Campbell-Smith, "A Viral Coyote-Badger Video Demonstrates the Incredible Complexity of Nature."

22   an aquatic version of this partnership: De Waal, *Are We Smart Enough to Know How Smart Animals Are?*, 199–200.

23   tool use all over the animal kingdom: many examples of animal tool use can be found in Shumaker, Walkup, and Beck, *Animal Tool Behavior*, Morell, *Animal Wise*, and De Waal, *Are We Smart Enough to Know How Smart Animals Are?*

23   pick up burning sticks in their beaks or talons: Bonta, "Intentional Fire-spreading by 'Firehawk' Raptors in Northern Australia."

23   Even pigeons will use boxes as stools: Cook and Fowler, " 'Insight' in Pigeons."

23   washing sweet potatoes before eating them: Hirata, "Sweet-Potato Washing."

24   distinct "ecotypes": Foote et al., "Genome-culture Coevolution Promotes Rapid Divergence of Killer Whale Ecotypes." See also the Supplementary Information.

24   but Darwin thought: *The Descent of Man, and Selection in Relation to Sex*, 61.

24   A 2013 study: Isden et al., "Performance in Cognitive and Problem-Solving Tasks in Male Spotted Bowerbirds Does not Correlate with Mating Success."

24   told journalist Virginia Morell: Morell, *Animal Wise*, 116.

25   researchers wrote of one such mother: Biro et al., "Chimpanzee Mothers at Bossou, Guinea, Carry the Mummified Remains of Their Dead Infants."

25   the orca J35, also known as Tahlequah: Marino, "The Harmful Effects of Captivity and Chronic Stress on the Well-Being of Orcas."

26   "zebrafishes will pay a cost": Balcombe, *What a Fish Knows*, 82.

27   "We hear no screams and see no tears": Balcombe, *What a Fish Knows*, 232.

CHAPTER 3: PHILOSOPHIES OF THE NON-HUMAN

28   A skull from a toddler *Australopithecus africanus*: Berger and McGraw, "Further Evidence for Eagle Predation of, and Feeding Damage on, the Taung Child."

28   A skull of one of our relatives called Paranthropus: Brain, "New Finds at the Swartkrans Australopithecine Site."

29   Fossil hominins some 1.8 million years old: Gibbons, "Meet the Frail, Small-brained People Who First Trekked out of Africa."

29 Being prey shaped us: This paragraph is based on Hart, *Man the Hunted.*

29 we were hunting animals for meat ourselves: Pobiner, "Meat-Eating Among the Earliest Humans."

29 the oldest figurative painting in the world: Aubert et al., "Earliest Hunting Scene in Prehistoric Art."

30 the mysterious antlered Celtic god Cernunnos: Fickett-Wilbar, "Cernunnos," 85–108.

30 Siip (or Sip), an old man, with a deer's antlers and ears: Looper, *The Beast Between,* 150.

30 possessors of spiritual power: Anderson, *Creatures of Empire,* 19–20.

30 The brown bears of Hokkaido: Isabella, "From Prejudice to Pride."

30 the ancient Greeks: My account is drawn from Clark, "Animals in Classical and Late Antique Philosophy."

31 *Fioretti di San Francesco*: Francis of Assisi, *The Little Flowers of Saint Francis of Assisi.*

32 and their skins peeled off of them: Joint Secretariat, *Inuvialuit and Nanuq,* 206.

32 the fascinating history of non-human animals put on trial: My account of these trials is drawn from Evans, *The Criminal Prosecution and Capital Punishment of Animals.*

33 "anathema and perpetual malediction": Ibid, 22

33 "to implore pardon for our sins": Ibid, 38–39.

33 "conferred upon them at the time of their creation": Ibid, 43.

33 citing Genesis: Genesis 1:30, KJV.

33 Genesis also said that man had dominion: Genesis 1:26, KJV.

34 "until death ensueth": Ibid, 143.

34 "had not participated in her master's crime of her own free-will": Ibid, 150.

35 a close reading of his own words suggests: Cottingham, " 'A Brute to the Brutes?' "

35 "That end is man": Quoted in Singer, *Animal liberation,* 203.

36 "but, *Can they suffer?*": Quoted in Regan, *The Case for Animal Rights,* 78.

36 In the nineteenth century: This paragraph and the next are based on Bates, "Have Animals Souls?"

36 an influential group of scientists known as behaviorists: Morell, *Animal Wise,* 13–14.

37 "The animal is not allowed to live before it dies": Quoted in Keeling et al., "Understanding Animal Welfare," 19–20.

38 "cannot involve this particular kind of loss": Singer, *Animal Liberation,* 21.

38 "the mystery of a unified psychological presence": Regan, *The Case for Animal Rights,* xvi.

39 "human chauvinism": Ibid, 31.

39 "Once we give up our claim to 'dominion'": Singer, *Animal Liberation,* 226.

39 "far more harm than good": Singer, *Animal Liberation,* 226.

39 "let them be": Regan, *The Case for Animal Rights,* xxxvii.

40 the famous "trolley problem": The original trolley problem was presented in Foot, "The Problem of Abortion and the Doctrine of Double Effect." The versions presented

here derive from the Bystander at the Switch and Fat Man cases in Thompson, "The Trolley Problem."

41 our "reflective intuitions": Regan, *The Case for Animal Rights*, 134.

41 In contrast, philosopher Lori Gruen: Gruen, *Entangled Empathy*.

41 "moral particularism": Dancy, "Moral Particularism."

41 She defines "entangled empathy": Gruen, *Entangled Empathy*, 3.

41 "it further entrenche[s] stereotypical gender roles": Ibid, 32.

42 a quicker, automatic emotional process and a slower, conscious, more rational process: Greene et al., "An fMRI Investigation of Emotional Engagement in Moral Judgment."

42 thinking in the switch-pulling case seems not to involve these areas: Ibid.

43 a journalist at the Atlantic asked one of the scientists who study moral cognition: Davis, "Do Emotions and Morality Mix?"

44 according to philosopher Rosalind Hursthouse: Hursthouse, "Virtue Ethics and the Treatment of Animals."

44 "mindful of others' rights to make their own choices": Ibid, 132.

45 "I continue to question whether the Court was right to deny leave": State of New York Court of Appeals, Motion No. 2018-268, 6–7.

45 a capuchin monkey throwing a cucumber in the face of a researcher: Brosnan and De Waal, "Monkeys Reject Unequal Pay."

CHAPTER 4: BETWEEN DOG AND WOLF

47 In July 2013 a young black wolf: Details of 47's collaring come from Landers, "Wolf 47 Works Full-Time for Washington Wildlife Researchers."

48 "Genetic integrity" is a phrase commonly used: Rohwer and Marris, "Is There a Prima Facie Duty to Preserve Genetic Integrity in Conservation Biology?"

50 most Italian "wolves" are wolf dogs of varying percentage: Randi et al., "Mitochondrial DNA Variability in Italian and East European Wolves."

52 "[c]oated in hair like the god of the animals": George, *The Epic of Gilgamesh*, 5.

52 "gazing at the lofty cedars": Ibid, 39.

54 I wrote one of them myself: Marris, *Rambunctious Garden*.

54 "where man himself is a visitor who does not remain": Public Law 88-577, "To Establish a National Wilderness Preservation System."

54 "independent of human design, control, and impacts": Sandler, *Environmental Ethics*, 42.

55 One study says that the timing of extinctions: Faurby et al., "Brain Expansion in Early Hominins Predicts Carnivore Extinctions in East Africa."

55 North America lost more than 70 percent of its megafauna: Broughton and Weitzel, "Population Reconstructions for Humans and Megafauna Suggest Mixed Causes for North American Pleistocene Extinctions."

56 These studies have shown: Gill et al., "Pleistocene Megafaunal Collapse, Novel Plant Communities, and Enhanced Fire Regimes in North America."

56 according to one analysis: Johnson, "Ecological Consequences of Late Quaternary extinctions of megafauna."

57 shade-adapted species took over: Ibid.

57 There were likely some secondary extinctions: Galetti et al., "Ecological and Evolutionary Legacy of Megafauna Extinctions."

57 lost seven entire genera of vultures: Perrig et al., "Demography of Avian Scavengers After Pleistocene Megafaunal Extinction."

57 The common vampire bat was likely hit hard: Galetti et al., "Ecological and Evolutionary Legacy of Megafauna Extinctions."

58 jaguars shrank to scale: Ibid.

58 As one scientific paper put it: Smith et al., "Unraveling the Consequences of the Terminal Pleistocene Megafauna Extinction on Mammal Community Assembly."

59 "With a long, long history of cultural use": Kimmerer, Braiding Sweetgrass, 164.

59 "It is highly likely that over centuries or perhaps millennia": Anderson, Tending the Wild, 156.

59 bringing fan palms to the Sonoran desert: Ibid, 155.

59 The Kumeyaay, who live in and around San Diego: Shipek, "A Native American Adaptation to Drought."

59 They brought prickly pear cactus from the desert: Harris et al., Foraging and Farming, 163–164.

59 anthropologist Florence Shipek writes: Shipek, "A Native American Adaptation to Drought."

60 Many Indigenous peoples use fire: Anderson, Tending the Wild, 148–155.

60 One theory about how dogs came to be domesticated: Pierotti and Fogg, The First Domestication.

60 Elephants are losing their tusks: Chiyo et al., "Illegal Tusk Harvest and the Decline of Tusk Size in the African Elephant."

60 are becoming smaller-bodied: Darimont et al., "Human Predators Outpace Other Agents of Trait Change in the Wild."

61 have shifted the pitch of their calls: Roca et al., "Shifting Song Frequencies in Response to Anthropogenic Noise."

61 Crows in Sendai, Japan: Nihei and Higuchi, "When and Where Did Crows Learn to Use Automobiles as Nutcrackers."

61 White storks in Spain stopped migrating: Gilbert, Movement and Foraging Ecology of Partially Migrant Birds in a Changing World.

61 540 million of them: Ackerman, The Genius of Birds, 242.

61 they have mastered automatic sliding doors: Ackerman, *The Genius of Birds*, 246.

61 learned to incorporate cigarette butts into their nests: Suárez-Rodríguez et al., "Incorporation of Cigarette Butts into Nests Reduces Nest Ectoparasite Load in Urban Birds."

61 laying eggs two weeks earlier: Ackerman, *The Genius of Birds*, 257.

61 evolving to lay their eggs on new plants: Singer and Parmesan, "Lethal Trap Created by Adaptive Evolutionary Response to an Exotic Resource."

61 1.2 miles a year: Platts et al., "Habitat Availability Explains Bariation in Climate-driven Range Shifts Across Multiple Taxonomic Groups."

61 Over half of plant and animal species: Weiskopf et al., "Climate Change Effects on Biodiversity, Ecosystems, Ecosystem Services, and Natural Resource Management in the United States."

62 scientists surveyed for birds in 1985 and in 2017: Freeman et al., "Climate Change Causes Upslope Shifts and Mountaintop Extirpations in a Tropical Bird Community."

62 shrinking by 0.035g per year: Prokosch, et al., "Are Animals Shrinking Due to Climate Change?"

62 The American lobster and Atlantic cod: Weiskopf et al., "Climate Change Effects on Biodiversity, Ecosystems, Ecosystem Services, and Natural Resource Management in the United States," 2.

63 was in many places *creating* those "natural" states: Bird and Nimmo, "Restore the Lost Ecological Functions of People."

63 When Christopher Columbus arrived: This statistic and the paragraph on "the Great Dying" that follows are derived from Koch et al., "Earth System Impacts of the European Arrival and Great Dying in the Americas after 1492."

63 "smaller shrubs and plants": Meany, *Vancouver's Discovery of Puget Sound*, 124.

63 where it was epidemic in 1775: This date and the 1782 date in the next sentence from Hopper, "Everyone Was Dead."

64 Brenda Guernsey, wrote in 2008: Guernsey, "Constructing the Wilderness and Clearing the Landscape," 112.

64 terra nullius—no one's land: Banner, "Why Terra Nullius?"

64 In California's Yosemite Valley: My account of Yosemite's origins is from Dowie, *Conservation Refugees*, 2–5.

64 "Staring in awe at the lengthy visas": Anderson, *Tending the Wild*, 3.

65 "Too dirty; too much bushy,": Taylor, "The Last Survivor."

65 "They could not even imagine": Kimmerer, *Braiding Sweetgrass*, 6.

66 the phrase "non-built environments": Sandler, *Environmental Ethics*, 42.

66 "spontaneity and otherness": Ibid, 44.

66 saving the Eastern Purple Martin from extinction: Jervis, et al., "Resisting Extinction."

CHAPTER 5: THE LION IN THE BACKYARD

70   "the judicious use of a small stick": Adamson, *Born Free*, 47–48.

70   "we became to some extent her prisoners": Ibid, 108.

71   McKenna said in a 2010 interview: "Elsa's Legacy: The Born Free Story," *Nature*.

71   "an exclusive adoption pack": Born Free Foundation, Adopt a Lion.

72   when investigative journalist Rachel Nuwer interviewed a woman: Nuwer, "The Strange and Dangerous World of America's Big Cat People."

74   an investigation by Sharon Guynup: Guynup, "Captive Tigers in the U.S. Outnumber Those in the Wild."

74   seems to cut across cultures: All the pets lists in this paragraph come from Hirschman, "Consumers and Their Animal Companions," 616–617.

75   Bieber was quoted as saying: Weaver, "Justin Bieber Would Like to Reintroduce Himself."

76   "act as extensions of the consumer's self": Hirschman, "Consumers and Their Animal Companions," 618.

76   When pop musician DJ Khaled posed: Noisey Staff, "DJ Khaled Reveals 'Major Key' Album Cover Which Is the Greatest Album Cover Maybe Ever."

77   as many as 40,000 years ago: Skoglund et al., "Ancient Wolf Genome Reveals an Early Divergence of Domestic Dog Ancestors and Admixture into High-Latitude Breeds."

77   when the ancient Gauls tied their female dogs to trees: Lescureux and Linnell, "Warring Brothers."

77   an article on wolf dogs as pets: Connors, "Do Wolfdogs Make Good Pets?"

78   Inyo, a wolf dog she raised from a pup: Terrill, *Part Wild*.

78   I pressed a dish towel to my bloody ear: Ibid, 209.

79   "and to leave it alone": Ibid, 239.

79   from three to nearly 400 square miles: Breton, "Influence of Enclosure Size on the Distances Covered and Paced By Captive Tigers."

79   up to 2,450 square miles: Mech, and Boitani, eds. *Wolves: Behavior, Ecology, and Conservation*, 21.

79   a 2006 story: Piovesan, "Wolf Dogs Killed Owner, Autopsy Determines."

79   and may actually instinctively fear us: Lazzaroni et al., "The Effect of Domestication and Experience on the Social Interaction of Dogs and Wolves with a Human Companion."

82   Philosopher Martha Nussbaum focuses on the "capabilities" of an individual: Nussbaum, *Frontiers of Justice*. Note that Nussbaum is working on a new book about animal justice, which she advised me should appear in 2022. Thus my summary may

not represent her more recent views. Readers who are interested in a justice approach to animal ethics are encouraged to seek out her forthcoming book when it appears.

82   "A society that does not guarantee these to all its citizens": Ibid, 75.

82   "sympathetic imagining": Ibid, 355.

82   "a chance to enjoy the light and air in tranquility": Ibid, 326.

83   "they live in splendid conditions": Adamson, *Born Free*, 39.

CHAPTER 6: THE AUTOCRATIC MENAGERIE

85   lion hunts undertaken by kings in Mesopotamia: Brereton, "Lion Hunting."

85   Ramses IX sent live hippos: Hancocks, *A Different Nature*, 7.

85   As philosopher Stephen R. L. Clark writes: "Animals in Classical and Late Antique Philosophy," 45.

85   The first modern zoo: The London Zoo material is drawn from Hancocks, *A Different Nature.*

86   Animal importer Carl Hagenbeck: Hagenbeck's story is drawn from Rothfels, *Savages and Beasts.*

86   "unadulterated people of nature": Ibid, 83.

86   a Mbuti man named Ota Benga: Delaney, "Ota Benga."

86   "comfortable, secure, and, of course, enlightened": Rothfels, *Savages and Beasts*, 183.

87   This conservation focus is now *mandatory*: Association of Zoos & Aquariums, "The Accreditation Standards & Related Policies," 24–25.

87   "human predation": Norton et al. eds. *Ethics on the Ark*, 46.

88   the telltale collapsed dorsal fin of the captive male orca: Blackfish showed SeaWorld employees claiming that 25% of orcas have a collapsed fin, but the real figure is much lower. The study I think they were referring to described an orca population in New Zealand in which 23.3% had "collapsing, collapsed, or bent" dorsal fins—most showing minor bends, twists, or droops. See Visser, "Prolific Body Scars and Collapsing Dorsal Fins on Killer Whales (*Orcinus orca*) in New Zealand Waters." This study itself explicitly noted that this rate was unusually high, noting that "For the British Columbian population the rate is 4.7% and for the Norwegian population the rate is 0.57%" and adds that "complete collapse of the dorsal fin of killer whales does not appear to be common in any population." In contrast, as of 2013, all male orcas in captivity had collapsed dorsal fins and some females also had collapsed fins according to Parsons et al., *An Introduction to Marine Mammal Biology and Conservation*, 168.

88   SeaWorld saw attendance crater: Mollman, "Attendance at SeaWorld San Diego Has Plummeted since the 'Blackfish' Documentary."

88 "We will increase our focus on rescue operations": Allen, Greg.

88 "connect in an inspiring new way": SeaWorld, "Orca Encounter."

89 would be at least 2.8 million square feet: The Whale Sanctuary Project, "We're building a model sanctuary."

89 SeaWorld's revenue in 2019: SeaWorld Entertainment. *2019 Annual Report.*

89 schoolchildren of Seattle donated their pennies: Bierlein, *Woodland*, 26.

89 Tusko: Pierce, "Tusko the Elephant Rampages through Sedro-Woolley."

89 "Elmer the Safety Elephant": Bierlein, Woodland, 59, 71.

89 the 2012 publication in the *Seattle Times* of "Glamour Beasts": Most of Chai's story and data on captive elephant mortality comes from Berens, "Glamour Beasts." A few details were sourced from elsewhere, and those will be noted separately.

90 overruled by the mayor: Hancocks, "Former Woodland Park Head Sees Sad Future for Elephants in OK."

90 50 miles a day: Siebert, "Zoos Called It a 'Rescue.' "

90 Hancocks reminisced in the *Seattle Times*: Hancocks, "Bamboo Should Be Sent to a Place Where She Can Heal."

90 "disciplined" with ax handles: Hancocks, "Former Woodland Park Head Sees Sad Future for Elephants in OK."

91 Hansa, was born in 2000: Mapes, "How Quickly They Grow."

91 1.2 million people came to see her: Bierlein, *Woodland*, 147.

91 According to Seattle's alternative newspaper: Madrid, "Cash Cows."

91 that training involved being hit with a bullhook: Hancocks, "Hansa's Short Life One of Deprivation."

91 Hansa was beaten to stop her from eating dirt: Hancocks, "Bamboo Should Be Sent to a Place Where She Can Heal."

91 cupcakes made of cornmeal: Mapes, "How Quickly They Grow."

92 Several of the staff wept: Mapes, "Seattle Zoo's Beloved Young Elephant Dies."

92 "She deeply touched our lives": Mulady, "Donation Made to Honor Hansa."

92 so the other elephants could mourn her: Madrid, "Cash Cows."

92 an editorial co-written by a member of the zoo's board: McGraw and Foster, "When You Stand Up for Zoos You Stand Up for Elephants."

92 The zoo's conservation director flew to Borneo: *Seattle Times* staff, "Conservation Gift to Honor Hansa."

92 6.5 million dollars in tickets and parking fees: Woodland Park Zoological Society, IRS form 990 for 2007.

92 an op ed from two members of their board of directors: Slinker and Liddell, "Op-ed: Zoos Play a Vital Role Protecting Wild Elephants and Their Habitat."

92   A medical report: I was unable to find the original report online and the Woodland
      Park Zoo did not respond to requests for comment of fact-checking assistance. I found
      the quotes passages in a blog post at the Zoo: Woodland Park Zoo, "Update: Learning
      More About Watoto."

93   On cue, Martin Ramirez, the zoo's curator of mammals, reminded the public:
      Ibid.

94   The *Seattle Times* found documents: Doughton,"Elephant Chai Suffered Injuries,
      Weight Loss Months Before Her Death."

94   about 300 elephants at AZA-accredited zoos: Siebert, "Zoos Called It a 'Rescue.'"

94   imported 17 elephants from a private game reserve: Details of this "rescue" are from
      Siebert, "Zoos Called It a 'Rescue.'"

94   tightened the rules on trading live African elephants: Winsor, "CITES Stops Short of
      Outright Ban on Zoo Trade in Wild African Elephants."

95   the AZA announced in 2019: American Veterinary Medical Association. "AZA to Phase
      Out Bullhooks for Elephant Management."

95   Researchers divide the odd behaviors in captive animals into two categories: Breton
      and Barrot, "Influence of Enclosure Size on the Distances Covered and Paced by
      Captive Tigers."

96   80 percent of zoo carnivores: Mason and Latham, "Can't Stop, Won't Stop."

96   64 percent of zoo chimps: Jacobson et al., "Characterizing Abnormal Behavior in a
      Large Population of Zoo-housed Chimpanzees."

96   85 percent of zoo elephants: Greco et al., "The Days and Nights of Zoo Elephants."

96   the Toledo Zoo has dosed zebras and wildebeest: Laidman, "Zoos Using Drugs to Help
      Manage Anxious Animals."

98   "They have a conception of freedom and a desire for it": Hribal, *Fear of the Animal
      Planet*, 26.

99   "Bob," "out," "key": Hess, *Nim Chimpsky*, 290.

99   "It's fine to get attached to the animals": Parker, "Killing Animals at the Zoo."

99   The National Elephant Center: Holsman, "New Details Emerge about Elephant Deaths
      at Fellsmere Center."

101   "more than 40 reintroduction programs": Association of Zoos & Aquariums, "Zoo
      and Aquarium Statistics."

101   800,000 animals: Ibid.

101   only 7 percent focus on species conservation: Loh, et al., "Quantifying the Contribution
      of Zoos and Aquariums to Peer-reviewed Scientific Research."

102   "The wolves are conservation ambassadors": Woodland Park Zoo. "Young Wolves
      Join Northern Trail."

103 200 million people visit a zoo every year: Association of Zoos & Aquariums, "Zoo and Aquarium Statistics."

103 Researchers quizzed visitors: Clayton et al., "The Role of Zoos in Fostering Environmental Identity."

103 A 2008 study of 206 zoo visitors: Clayton et al., "Zoo Experiences."

104 check out his dissertation: Fraser, "An Examination of Environmental Collective Identity Development Across Three Life-Stages."

CHAPTER 7: THE DIGNITY OF THE CONDOR

110 killed off seven other kinds of vultures: Perrig et al., "Demography of Avian Scavengers after Pleistocene Megafaunal Extinction."

110 Native people up and down the coast: Foster, "Wings of the Spirit."

110 prey-go-neesh: Smith, Anna, "An Indigenous Effort to Return Condors to the Pacific Northwest Nears Its Goal."

110 Its range had shrunk to just the San Joaquin Valley: Meretsky et al., "Demography of the California Condor."

110 strychnine and other poisons: Snyder and Snyder, "Biology and Conservation of the California Condor."

111 "*hasten* the extinction of the California condor": Phillips and Nash, eds., *The Condor Question*, 12.

111 "much more than feathers, flesh, and genes": Ibid, 45, 60.

111 "the right to their own dignity": Ibid, 273.

111 "die with the dignity that has always been yours": Stallcup, "Farewell, Skymaster."

111 sued to stop the recovery team: *National Audubon Society v. Hester.*

112 "a scientific twilight zone": *Sports Illustrated* Staff, "Last Chance for the Condor."

112 "the condor was destroyed in order to save it": Ibid.

112 causing the extinction of the California condor louse: Jørgensen, "Conservation Implications of Parasite Co-reintroduction."

113 a 1910 report: Finley, "Life History of the California Condor Part IV."

115 As of March 2020: San Diego Zoo Global Library, California Condor (*Gymnogyps californianus*) Fact Sheet

115 "look for the number printed on its wing tags": U.S. Fish & Wildlife Service, "California Condor Population Information."

115 76 recorded condor deaths since 1992: Mock, "Lead Ammo, the Top Threat to Condors, Is Now Outlawed in California."

116 "He carries our prayers to the heavens": Freeman, "Return of the Sacred Condor."

116 "his many contributions to California condor recovery": The Condor Cave, "The Loss of AC-9."

117 "The birds I observed in Sierra San Pedro Martir": Farnsworth, "The Condor Question Revisited."

118 some 2,300 northern white rhinoceroses lived in the wild: Scott et al., *Shepherding Nature*, 42.

120 defined the proper objective of conservation biology: Soulé, "What Is Conservation Biology?"

121 a scientific paper in the journal *Conservation Biology*: Meretsky, et al., "Demography of the California Condor: Implications for Reestablishment."

CHAPTER 8: ARE SPECIES VALUABLE?

123 "The inherent value and rights of individuals do not wax or wane": Regan, *The Case for Animal Rights*, xxxix.

123 "if species were becoming extinct in ways that had no impact": Nussbaum, *Frontiers of Justice*, 357.

123 "Thinking Like a Mountain": Leopold, *A Sand County Almanac*, 129–133.

124 "The Land Ethic": Ibid, 201–226.

124 "solitary, poor, nasty, brutish, and short": Lloyd and Sreedhar, "Hobbes's Moral and Political Philosophy."

125 an influential 1967 article: White, "The Historical Roots of Our Ecologic Crisis."

126 a distinction between instrumental value and final value: my taxonomy of value is derived from Sandler, *The Ethics of Species*.

126 "Objective value is discovered by valuers": Ibid, 18.

127 Imagine that the last person on Earth: Routley, "Is There a Need for a New, an Environmental, Ethic?"

127 "[T]he valuing might be for what the entity represents": Sandler, *The Ethics of Species*, 18–19.

128 "Whether the blue whale survives": Routley, "Is There a Need for a New, an Environmental, Ethic?"

128 "Someone might be outraged by the practice of euthanizing": Sandler, *The Ethics of Species*, 71.

129 "species are too diffuse": Ibid, 38.

130 "shut down a story of many millennia": Bekoff, Marc, ed., *Encyclopedia of Animal Rights and Animal Welfare*, 206.

130 "the absence of reasons to believe": Sandler, *The Ethics of Species*, 33.

131 "the variety of life on Earth at all its levels": American Museum of Natural History. "What Is Biodiversity?"

132 is older than about 12,000 years: Jackson, "Perspective: Ecological Novelty Is Not New."

133 Darwin predicted: Arditti et al., "'Good Heavens What Insect Can Suck It?'"

133 One of my favorite studies from the last few years: Schilthuizen et al., "Incorporation of an Invasive Plant into a Native Insect Herbivore Food Web."

134 "Come back in a million years": Thomas, *Inheritors of the Earth*, 179.

134 "the process of intentionally altering a site": Higgs, *Nature by Design*, 107.

134 the role of history in defining goals is being more openly questioned: Rohwer and Marris, "Renaming Restoration."

CHAPTER 9: FEEDING POLAR BEARS

137 the rate is an impressive 75 percent: Smith, "Winter Bird Feeding."

137 and boost the number of chicks that fledge: Robb et al., "Winter Feeding of Birds Increases Productivity in the Subsequent Breeding Season."

137 The stay-at-home birds are evolving longer beaks: Plummer, "Is Supplementary Feeding in Gardens a Driver of Evolutionary Change in a Migratory Bird Species?"

138 "How can you *not* touch the whale": Chisholm and Parfit, *Saving Luna*.

139 "Supplementary feeding isn't advised": Parrott, "How You Can Help—Not Harm—Wild Animals Recovering from Bushfires."

139 "Wildlife are highly resourceful": Santa Monica Mtns, Twitter Post.

139 the clip was watched 2.5 billion times: Mittermeier, "Starving-Polar-Bear Photographer Recalls What Went Wrong."

139 "Of course, that crossed my mind": Gibbens, "Heart-Wrenching Video Shows Starving Polar Bear on Iceless Land."

139 "climate warming poses the single most important threat": Wiig et al., "*Ursus maritimus.*"

139 A study of nine female polar bears: Pagano et al., "High-energy, High-fat Lifestyle Challenges an Arctic Apex Predator."

140 shrinking by 14 percent every decade: Ibid.

140 a 2015 report: Joint Secretariat, *Inuvialuit and Nanuq*.

140 "fewer really big bears": Ibid, xiv.

140 26,000 polar bears in the world: Regehr et al., "Conservation Status of Polar Bears."

141 reduce the number of bears by about 30 percent by 2050: Ibid.

141 uses the polar bear as a case study: Much of the rest of this chapter works through the arguments in Palmer, "Should We Provide the Bear Necessities?"

141   "the laissez-faire intuition": See Palmer, *Animal Ethics in Context.*

143   took home nearly 8 million dollars in 2019: Nicklaus, "Peabody CEO Earns $7.6 Million."

145   12,325 calories a day: Pagano et al., "High-energy, High-fat Lifestyle Challenges an Arctic Apex Predator."

146   according to philosophers Sue Donaldson and Will Kymlicka: Donaldson and Kymlicka, *Zoopolis.*

146   "but as visitors to foreign lands": Ibid, 170.

146   "Animals have evolved to survive under these conditions": Ibid, 182.

146   "allows a community to get back on its own feet": Ibid, 181.

146   "where there is no way to restore habitat": Mathews, "Wild Animals Are Starving, and It's Our Fault, so Should We Feed Them?"

148   "A species is conservation reliant if ": Scott et al., *Shepherding Nature,* 3.

148   84 percent require ongoing management: Ibid, 91–97.

148   The Kihansi spray toad: Ibid, 84.

149   The Oregon silverspot butterfly: Ibid, 36.

149   Grevy's zebras: Marris, "These Rare Zebras Are Dependent on Humans."

149   The southern white rhino: Scott et al., *Shepherding Nature,* 30–32.

149   throwing these cousins into proximity: Smart, "Grizzly Bears Move North in High Arctic as Climate Change Expands Range."

149   split less than 500,000 years ago: Rinker et al., "Polar bear evolution is marked by rapid changes in gene copy number in response to dietary shift."

150   scientists found that the hybrids actually have higher survival rates: Fitzpatrick and Shaffer, "Hybrid Vigor Between Native and Introduced Salamanders."

CHAPTER 10: THE ARROW'S TIP

152   Elias Machipango Shuverireni: Note that the Matsigenka use the Spanish patronymic followed by mother's maiden name, so his surname in the English sense is Machipango.

154   Snakes are the arrows of invisible hunters: Shepard, "Hunting in Amazonia."

158   a complex and often bloody history: Shepard et al., "Trouble in Paradise."

159   "Those strange white conquerors": Hecht, *The Scramble for the Amazon,* 437.

162   "but, rather, as a long-term social relationship": Nadasdy, *Sovereignty's Entailments,* 269.

162   "Hunting was not only a display of human prowess": Anderson, *Creatures of Empire,* 28.

162   "negotiation with prey animals' spiritual protectors": Ibid, 30.

162 "mutual support": Ibid, 31.

162 a ritual called Loojil Ts'oon: Santos-Fita et al., "Symbolism and Ritual Practices Related to Hunting in Maya Communities."

163 "Killing a *who* demands something different": Kimmerer, *Braiding Sweetgrass*, 183.

166 "aggressively ethnocentric": Plumwood, "Integrating Ethical Frameworks for Animals, Humans, and Nature."

166 inspired by being attacked by a saltwater crocodile: Plumwood, "Human Vulnerability and the Experience of Being Prey."

167 "it has seemed to me that our worldview denies the most basic feature of animal existence": Plumwood, "Tasteless."

167 "systems of flow and exchange that nurture all life": Plumwood, "Integrating Ethical Frameworks for Animals, Humans, and Nature."

CHAPTER 11: BLOODSHED FOR BIODIVERSITY

169 Karl Campbell is a middle-aged, medium-sized Australian: Karl's story has been adapted from Marris, "Process of Elimination."

169 the Galápagos Islands,: Excellent overviews of the history of and conservation dilemmas in the Galápagos can be found in Hennessy, *On the Backs of Tortoises*, and Nicholls, *The Galápagos*.

170 The roughly 465,000 islands in the world: Holmes et al., "Globally Important Islands."

170 reuniting Pangea in practice: Thomas, *Inheritors of the Earth*.

171 in 1994, a biological expedition: Marris, "Resurrecting a Long-Lost Galapagos Giant Tortoise."

172 a single cat was found to be responsible for eating seven endangered Galápagos penguins every month: Steinfurth, "Marine Ecology and Conservation of the Galápagos Penguin."

172 62 percent of amphibian, reptile, bird, and mammal extinctions: Bellard, "Alien Species as a Driver of Recent Extinctions."

173 kill up to 4 billion birds and 22.3 billion mammals: Loss et al., "The Impact of Free-Ranging Domestic Cats on Wildlife of the United States."

173 Cats have been a factor in 63 extinctions/all the species driven extinct since 1500: My source for this is Doherty et al., "Invasive Predators and Global Biodiversity Loss." In particular, I downloaded the table of raw data from the Supplemental material, pulled out the extinct species, and then looked at their geographical locations.

173 They've been slowly ambling north ever since: Walsh and Tucker, "Contemporary Range Expansion of the Virginia Opossum."

174 arrived there less than 20,000 years ago: Thomas, *Inheritors of the Earth*, 84.

174 Arctic ptarmigan used to roam Central Europe: Ibid, 85.

174 domesticated about 2,000 years ago: Orlando, "Back to the Roots and Routes of Dromedary Domestication."

174 roam free in the Outback: Lundgren et al., "Introduced Megafauna are Rewilding the Anthropocene."

174 beavers are moving into the Arctic tundra: Tape, Ken D et al. "Tundra Be Dammed."

176 One particularly massive operation: Marris, "Large Island Declared Rat-Free in Biggest Removal Success."

176 the "death row dingoes" of Pelorus Island: This moniker comes from Schwartz, "Death row dingoes set to be the environmental saviour of Great Barrier Reef's Pelorus Island." Details of the project are from Allen et al., "Elucidating Dingo's Ecological Roles."

176 "I do not believe this step is necessary or desirable": Soulé, "What Is Conservation Biology?"

177 211,560 cats in twelve months: Doherty et al., "Conservation or Politics?"

177 Also in 2015: I pulled this list from The Database of Island Invasive Species Eradications, developed by Island Conservation.

177 "a being which cannot see itself as an entity with a future": Quoted in Norcross, "Death for Animals," 468.

177 Others argue that even if some animals don't prefer to exist: Ibid, 466.

178 brodifacoum: Littin et al., "Comparative Effects of Brodifacoum on Rats and Possums."

180 lose as much as 40 percent of their crop to rodents: According to Karl Campbell, this figure is derived from an unpublished study by the Ecuadorian Ministry of Agriculture.

182 Le Mitouard, Eric. "Sa pétition pour sauver les rats de Paris a déjà recueilli 17,000 signatures." Updated number of signatories from https://www.mesopinions.com/petition/animaux/stoppez-genocide-rats/26805

183 the government of Alberta has shot more than 1,000 wolves: Marris, "Wolf Cull Will Not Save Threatened Canadian Caribou."

183 Cowbirds evolved this trick: Daley, "North America's Rarest Warbler Comes Off the Endangered List."

183 an unofficial military strategy against the continent's Indigenous people: Phippen, "'Kill Every Buffalo You Can!'"

183 so closely related to spotted owls: Haig, "Genetic Identification of Spotted Owls, Barred Owls, and Their Hybrids."

183 Barred owls can live more places: Clément, "Habitat Features and Behavioral Plasticity Promote Barred Owl Presence in Developed Landscapes."

184 3,135 owls: U.S. Fish & Wildlife Service, "Barred Owl Study Update." Accessed November 13, 2020.

184 hybridization was automatically assumed to be a threat: Mallet et al., "How Reticulated Are Species?"

184 10–30 percent of multicellular animal and plant species: Abbott et al., "Hybridization and speciation."

185 The Genovesa mockingbird: Nietlisbach et al., "Hybrid Ancestry of an Island Subspecies of Galápagos Mockingbird."

185 We ourselves are hybrids: Ackermann et al., "The Hybrid Origin of 'Modern' Humans."

186 Ferns somehow copied a gene from a mosslike species of hornwort: Li et al., "Horizontal Transfer of an Adaptive Chimeric Photoreceptor."

CHAPTER 12: THE FRIENDLY TOUTOUWAI

189 Aotearoa broke off from the mega-continent: For background on the geology, ecology, and environmental history of Aotearoa and the origins and impacts of its imported mammal species, I relied on King, *The Handbook of New Zealand Mammals*.

190 less than 1,000 years ago: Wilmshurst et al., "High-precision Radiocarbon Dating Shows Recent and Rapid Initial Human Colonization of East Polynesia."

190 *Hoplodactylus delcourti*: Bauer and Russell, "*Hoplodactylus delcourti*."

191 The IUCN listing: Hitchmough et al., "*Hoplodactylus delcourti*."

191 53 species of birds since human arrival: Miskelly, "Extinct Birds of New Zealand."

191 41 percent of New Zealanders own a pet cat: Companion Animals New Zealand. Companion Animals in New Zealand 2020.

192 "You didn't see that one coming, did you!": Predator Free NZ, "Backyard Trapping."

192 a possum hunting contest: Newshub reporter, "Baby Possums Drowned."

192 dressed up in costumes: Smith, "Possum Pics Disgust."

192 "If you see a possum on the road": NZ Pocket Guide, "Why New Zealand Hates Possums."

192 "One aerial application can kill over 95% of possums": Forest & Bird, "Frequently Asked Questions about 1080."

193 the associated suffering is rated as merely "severe": Littin et al., "Welfare Aspects of Vertebrate Pest Control and Culling."

193 "these substances cause such intense and prolonged suffering": Royal New Zealand Society for the Prevention of Cruelty to Animals Incorporated. "1080—What Is It, and What Can Be Done About It?"

193 a woman from Pukekohe: Royal New Zealand Society for the Prevention of Cruelty to Animals Incorporated, "SPCA Prosecutes Woman for Neglecting Her Pet Rabbit."

193   the government dropped 1080 on 22,000 hectares: Auckland Council, "Hunua Pest Numbers at Record Low."

194   minor-key song: personal communication with musician Theodore Davis.

194   1,400 breeding pairs of North Island kōkako: BirdLife International, *Callaeas wilsoni.*

194   Kapiti Island: Historical background from Department of Conservation (NZ). "History of Kapiti Island." Animal control details from Department of Conservation (NZ). "Kapiti Island Nature and Conservation."

195   a combination of rats and loss of forest: BirdLife International, *Notiomystis cincta.*

196   "N/A due to mixed provenance": Brown et al., *Kapiti Island Ecological Restoration Strategy.*

196   lost its takahē in 1894: BirdLife International, *Porphyrio mantelli.*

197   Eleven thousand school kids: Zealandia. *Annual Report 2018/19*, 6.

199   Ecologist Chris Thomas says: Thomas, *Inheritors of the Earth*, 124–132.

200   a critique of the campaign in the journal *Conservation Letters*: Linklater and Steer, "Predator Free 2050: A flawed Conservation Policy Displaces Higher Priorities and Better, Evidence-based Alternatives."

200   it might cost NZ $32 billion: Ibid.

201   The Treaty of Waitangi: Ministry for Culture and Heritage (NZ). "The Treaty in Brief."

202   NZ$2.272 billion in 90 separate settlements: Driver, "Fact Check."

202   1998 settlement with the Ngāi Tahu: Ministry for Culture and Heritage (NZ), "The Ngāi Tahu Claim."

CHAPTER 13: COMPASSIONATE CONSERVATION

206   two million open mine shafts: Pedler, "The Impacts of Abandoned Mining Shafts."

207   an interview with *Undark* magazine: Lewis, "An Identity Crisis for the Australian Dingo."

207   discovered Australia 65,000 years ago: Clarkson et al., "Human Occupation of Northern Australia by 65,000 Years Ago."

207   from Southeast Asia some 3,500 years ago: Balme et al., "New Dates on Dingo Bones from Madura Cave."

207   they step in their own footprints, like cats: Smith, *The Dingo Debate*, 17.

207   "little hungry wolves": Quoted in ibid, 3.

207   they "accept humans as social companions": Ibid, 20.

207   even breastfed by women: Philip, "Living Blanket, Water Diviner, Wild Pet."

210 Wallach and Lundgren collaborated with Wayne Linklater and Jamie Steer: Wallach et al., "When All Life Counts in Conservation."

211 In a 2018 paper: Wallach, "Summoning Compassion to Address the Challenges of Conservation."

211 fewer than 10 birds: Waters, "Knights in Shining Fur."

212 Foxes killed 142 penguins: Middle Island Project, "2017–2018 Penguin Breeding Season."

214 drive more 25 species extinct: Radford et al., "Degrees of Population-level Susceptibility of Australian Terrestrial Non-volant Mammal Species."

214 One academic effort to get a handle on the scale of control: Reddiex, "Control of Pest Mammals for Biodiversity Protection in Australia."

214 they have killed an estimated 844,000 cats: Department of the Environment and Energy (Australia), *Threatened Species Strategy—Year Three Progress Report*, 18.

216 Other ecologists say that this optimism simply isn't supported: Callen et al., "Envisioning the Future with 'Compassionate Conservation.'"

218 I challenged Wallach with the problem of Gough Island: Rohwer and Marris, "Clarifying Compassionate Conservation with Hypotheticals."

218 very closely related to the more widespread Wandering Albatross: Burg and Croxall, "Global Population Structure and Taxonomy of the Wandering Albatross Species Complex."

218 came to the island with people before 1888: Cuthbert, et al., "Population Trends and Breeding Success of Albatrosses and Giant Petrels at Gough Island."

219 wrote 35 conservation biologists and ecologists: Hayward et al., "Deconstructing Compassionate Conservation."

219 A 2020 paper with 36 authors: Callen et al., "Envisioning the Future with 'Compassionate Conservation.'"

CHAPTER 14: BILBY THUNDERDOME

221 "docility, as a result of becoming accustomed to humans for food rewards": Montague et al., "Comparative Analysis of the Domestic Cat Genome."

222 One look at "prey naiveté" around the world: Anton et al., "Global Determinants of Prey Naiveté to Exotic Predators."

227 put gene drives and CRISPR together: Esvelt et al., "Emerging Technology."

230 He calls one approach a "daisy chain": Noble et al., "Daisy-Chain Gene Drives for the Alteration of Local Populations."

230 He called the basic version of a gene drive unsuitable: Esvelt and Gemmell, "Conservation Demands Safe Gene Drive."

231 a project to disrupt the cycle of Lyme disease in Nantucket: Buchthal et al., "Mice Against Ticks."

231 Imagine making the bilby and the bettong smarter overnight: Rohwer, "A Duty to Cognitively Enhance Animals."

232 according to philosopher Ronald Sandler: Sandler, "The Ethics of Genetic Engineering and Gene Drives in Conservation."

232 A study of 8,199 New Zealanders: MacDonald et al., "Public Opinion Towards Gene Drive."

233 "Before, they had been beasts.": Wells, *The Island of Doctor Moreau*, 99.

234 is there any ethical reason to resurrect lost species like the mammoth?: Rohwer and Marris, "An Analysis of Potential Ethical Justifications for Mammoth De-extinction."

235 forced to live in a world without mammoths: Rohwer and Marris, "An Analysis of Potential Ethical Justifications for Mammoth De-extinction."

236 will kill a female's infants: Zipple et al., "Conditional Fetal and Infant Killing by Male Baboons."

237 In a now-famous study: Sapolsky and Share, "A Pacific Culture Among Wild Baboons."

237 Lori Gruen tells a story: Gruen, *Entangled Empathy*, 58.

238 "The conception of flourishing is thoroughly evaluative and ethical": Nussbaum, *Frontiers of Justice*, 366.

238 "the beginning, not the end, of evaluation": Ibid, 369.

239 "benevolent despotism": Ibid, 373.

239 "the question must remain a very difficult one": Ibid, 379.

239 "Animal cultures are full of humiliation": Ibid, 399.

239 "the gradual supplanting of the natural by the just": Ibid, 400.

239 "The wolf also shall dwell with the lamb": Isaiah 11:6–7, KJV.

240 "most sentient animals who come into existence": Horta, "Animal Suffering in Nature."

240 "vast, unceasing slaughter": McMahan, "The Meat Eaters."

241 Nussbaum invokes it when arguing: Nussbaum, *Frontiers of Justice*, 373.

242 seems ethically acceptable, even good: Rohwer, "Gene Drives, Species, and Compassion for Individuals."

242 The International Commission on the Clinical Use of Human Germline Genome Editing released a report: National Academy of Sciences 2020, Heritable Human Genome Editing.

242 Scientist He Jiankui was jailed: Cohen, "Commission Charts Narrow Path."

242 Some 72 percent of Americans: Pew Research Center, "Public Views of Gene Editing for Babies."

243 for mutual benefit, all together, with care and love: Kofler et al., "Editing Nature."

CHAPTER 15: HOW TO BE A GOOD HUMAN TO
THE NON-HUMAN WORLD

245 a paper by two Dutch scholars: Lenders and Janssen, "The Grass Snake and the Basilisk."

245 Lithuanians and Latvians did indeed let grass snakes live in their houses: Eckert, "On the Cult of the Snake in Ancient Baltic and Slavic Tradition."

246 A Romanian naturalist writes: Gheorghe, "The Grass Snake."

247 Just a quarter of the Earth's ice-free land: Riggio et al., "Global Human Influence Maps."

247 Thirty-eight percent of the Earth's ice-free land: Food and Agriculture Organization, "Land Use in Agriculture by the Numbers."

249 less than 15 percent of the land was transformed: Winter, "A Hawaiian Renaissance That Could Save the World."

251 "until a maggot dropped from his nostril": George, *The Epic of Gilgamesh*, 81.

251 Gilgamesh becomes overcome with fear: Ibid, 70.

252 Val Plumwood says that the two dueling modern Western ideas about death are both unsatisfying: Plumwood, "Tasteless," 323–330.

252 according to the *Sydney Morning Herald*: "Val Plumwood died of natural causes."

252 "a fountain of energy flowing": Leopold, *A Sand County Almanac*, 216.

253 Plumwood grappled with this tension twenty years ago: Plumwood, "Integrating Ethical Frameworks," 317.

256 some individual plants are more willing to take risks: Reed-Guy, "Sensitive Plant (*Mimosa pudica*) Hiding Time Depends on Individual and State."

257 I am a "value pluralist": Mason, "Value Pluralism."

259 Recently, Arian Wallach wrote a paper on moral residue: Batavia et al., "The Moral Residue of Conservation."

BIBLIOGRAPHY

Abbott, Richard, Dirk Albach, Stephen Ansell, Jan W. Arntzen, Stuart J. E. Baird, Nicolas Bierne, Jenny Boughman, et al. "Hybridization and Speciation." *Journal of Evolutionary Biology* 26, no. 2 (2013): 229–246.

Ackerman, Jennifer. *The Genius of Birds.* New York: Penguin, 2016.

Ackermann, Rebecca Rogers, Alex Mackay, and Michael L. Arnold. "The Hybrid Origin of 'Modern' Humans." *Evolutionary Biology* 43, no. 1 (2016): 1–11.

Adamson, Joy. *Born Free: A Lioness of Two Worlds.* United Kingdom: Fontana/Collins, 1960.

Allen, Benjamin L., Lee R. Allen, Michael Graham, and Matt Buckman. "Elucidating Dingo's Ecological Roles: Contributions from the Pelorus Island Feral Goat Biocontrol Project." *Australian Zoologist* (2020): 1.

Allen, Greg. "SeaWorld Agrees to End Captive Breeding of Killer Whales." *NPR*, March 17, 2016. https://www.npr.org/sections/thetwo-way/2016/03/17/470720804/seaworld -agrees-to-end-captive-breeding-of-killer-whales.

American Museum of Natural History. "What Is Biodiversity?" Accessed November 11, 2020. https://www.amnh.org/research/center-for-biodiversity-conservation/what-is -biodiversity.

American Veterinary Medical Association. "AZA to Phase Out Bullhooks for Elephant Management." JAVMA News. October 30, 2019. https://www.avma.org/javma -news/2019-11-15/aza-phase-out-bullhooks-elephant-management.

Anderson, M. Kat. *Tending the Wild: Native American Knowledge and the Management of California's Natural Resources.* Berkeley and Los Angeles: University of California Press, 2013.

Anderson, Virginia DeJohn. *Creatures of Empire: How Domestic Animals Transformed Early America*. New York: Oxford University Press, 2004.

Anton, Andrea, Nathan R. Geraldi, Anthony Ricciardi, and Jaimie T. A. Dick. "Global Determinants of Prey Naiveté to Exotic Predators." *Proceedings of the Royal Society B* 287, no. 1928 (2020): 20192978.

Appleby, Michael, Anna Olsson, and Francisco Galindo, eds. *Animal Welfare*, 3rd ed. Wallingford, Oxfordshire: CABI, 2018.

Arditti, Joseph, John Elliott, Ian J. Kitching, and Lutz T. Wasserthal. " 'Good Heavens What Insect Can Suck It'—Charles Darwin, *Angraecum sesquipedale* and *Xanthopan morganii praedicta*." *Botanical Journal of the Linnean Society* 169, no. 3 (2012): 403–432.

Association of Zoos & Aquariums. "The Accreditation Standards & Related Policies," 2nd ed., 2020.

Association of Zoos & Aquariums. "Zoo and Aquarium Statistics." Last updated April 2020. Accessed November 11, 2020. https://www.aza.org/zoo-and-aquarium-statistics ?locale=en.

Aubert, Maxime, Rustan Lebe, Adhi Agus Oktaviana, Muhammad Tang, Basran Burhan, Andi Jusdi, Budianto Hakim, et al. "Earliest Hunting Scene in Prehistoric Art." *Nature* 576, no. 7787 (2019): 442–445.

Auckland Council. "Hunua Pest Numbers at Record Low." Our Auckland. February 26, 2019. https://ourauckland.aucklandcouncil.govt.nz/articles/news/2019/02/hunua-pest -numbers-at-record-low/.

Balcombe, Jonathan. *What a Fish Knows: The Inner Lives of Our Underwater Cousins*. New York: Scientific American/Farrar, Straus and Giroux, 2016.

Balme, Jane, Sue O'Connor, and Stewart Fallon. "New Dates on Dingo Bones from Madura Cave Provide Oldest Firm Evidence for Arrival of the Species in Australia." *Scientific Reports* 8, no. 1 (2018): 1–6.

Banner, Stuart. "Why Terra Nullius? Anthropology and Property Law in Early Australia." *Law and History Review* 23, no. 1 (2005): 95–131.

Barnosky, Anthony D., Emily L. Lindsey, Natalia A. Villavicencio, Enrique Bostelmann, Elizabeth A. Hadly, James Wanket, and Charles R. Marshall. "Variable Impact of Late-Quaternary Megafaunal Extinction in Causing Ecological State Shifts in North and

South America." *Proceedings of the National Academy of Sciences* 113, no. 4 (2016): 856–861.

Bartal, Inbal Ben-Ami, Jean Decety, and Peggy Mason. "Empathy and Pro-social Behavior in Rats." *Science* 334, no. 6061 (2011): 1427–1430.

Bartal, Inbal Ben-Ami, Haozhe Shan, Nora MR Molasky, Teresa M. Murray, Jasper Z. Williams, Jean Decety, and Peggy Mason. "Anxiolytic Treatment Impairs Helping Behavior in Rats." *Frontiers in Psychology* 7 (2016): 850.

Batavia, Chelsea, Michael Paul Nelson, and Arian D. Wallach. "The Moral Residue of Conservation." *Conservation Biology* (2020).

Bates, A. W. H. "Have Animals Souls? The Late-Nineteenth Century Spiritual Revival and Animal Welfare." In *Anti-Vivisection and the Profession of Medicine in Britain*, 43–67. London: Palgrave Macmillan, 2017.

Bauer, Aaron M., and Anthony P. Russell. "*Hoplodactylus delcourti* n. sp. (Reptilia: Gekkonidae), the Largest Known Gecko." *New Zealand Journal of Zoology* 13, no. 1 (1986): 141–148.

Beiler, Kevin J., Daniel M. Durall, Suzanne W. Simard, Sheri A. Maxwell, and Annette M. Kretzer. "Architecture of the wood-wide web: Rhizopogon spp. genets link multiple Douglas-fir cohorts." *New Phytologist* 185, no. 2 (2010): 543–553.

Bekoff, Marc, ed. *Encyclopedia of Animal Rights and Animal Welfare*. 2 vols. Santa Barbara: ABC-CLIO, 2009.

Bellard, Céline, Phillip Cassey, and Tim M. Blackburn. "Alien Species as a Driver of Recent Extinctions." *Biology Letters* 12, no. 2 (2016): 20150623.

Berens, Michael J. "Glamour Beasts: The Dark Side of Elephant Captivity," *Seattle Times*. December 1, 2012. Part 1 at https://special.seattletimes.com/o/html/nationworld /2019809167_elephants02m.html.

Berger, L. R., and W. S. McGraw. "Further Evidence for Eagle Predation of, and Feeding Damage on, the Taung Child." *South African Journal of Science* 103, no. 11-12 (2007): 496–498.

Bierlein, John. *Woodland: The Story of the Animals and People of Woodland Park Zoo*. History Ink, 2017.

Bird, Rebecca Bliege, and Dale Nimmo. "Restore the Lost Ecological Functions of People." *Nature Ecology and Evolution* 2, no. 7 (2018): 1050–1052.

BirdLife International. 2016. Porphyrio mantelli. The IUCN Red List of Threatened Species 2016: e.T22728833A94998264. https://dx.doi.org/10.2305/IUCN.UK.2016-3.RLTS. T22728833A94998264.en. Downloaded on 13 November 2020.

BirdLife International. 2017. Callaeas wilsoni (amended version of 2016 assessment). The IUCN Red List of Threatened Species 2017: e.T103730482A119551156. https://dx.doi .org/10.2305/IUCN.UK.2017-3.RLTS.T103730482A119551156.en. Downloaded on 13 November 2020.

BirdLife International. 2017. Notiomystis cincta (amended version of 2016 assessment). The IUCN Red List of Threatened Species 2017: e.T22704154A118814893. https://dx.doi .org/10.2305/IUCN.UK.2017-3.RLTS.T22704154A118814893.en. Downloaded on 20 October 2020.

Biro, Dora, Tatyana Humle, Kathelijne Koops, Claudia Sousa, Misato Hayashi, and Tetsuro Matsuzawa. "Chimpanzee Mothers at Bossou, Guinea, Carry the Mummified Remains of Their Dead Infants." *Current Biology* 20, no. 8 (2010): R351–R352.

Bonta, Mark, Robert Gosford, Dick Eussen, Nathan Ferguson, Erana Loveless, and Maxwell Witwer. "Intentional Fire-spreading by 'Firehawk' Raptors in Northern Australia." *Journal of Ethnobiology* 37, no. 4 (2017): 700–718.

Born Free Foundation. "Adopt a Lion." Accessed December 1, 2020. https://www.bornfree .org.uk/adopt-a-lion.

Brain, Charles K. "New Finds at the Swartkrans Australopithecine Site." *Nature* 225, no. 5238 (1970): 1112–1119.

Brereton, Gareth. "Lion Hunting: The Sport of Kings." British Museum Blog, January 4, 2019. https://blog.britishmuseum.org/lion-hunting-the-sport-of-kings-2/.

Breton, Grégory, and Salomé Barrot. "Influence of Enclosure Size on the Distances Covered and Paced by Captive Tigers (*Panthera tigris*)." *Applied Animal Behaviour Science* 154 (2014): 66–75.

Brosnan, Sarah F., and Frans B. M. De Waal. "Monkeys Reject Unequal Pay." *Nature* 425, no. 6955 (2003): 297–299.

Broughton, Jack M., and Elic M. Weitzel. "Population Reconstructions for Humans and Megafauna Suggest Mixed Causes for North American Pleistocene Extinctions." *Nature Communications* 9, no. 1 (2018): 1–12.

Brown, Kerry, Jeremy Rolfe, Lynn Adams, Peter de Lange, and Chris Green. *Kapiti Island Ecological Restoration Strategy.* Wellington: Department of Conservation Technical Report, 2016.

Browning, Heather. "No Room at the Zoo: Management Euthanasia and Animal Welfare." *Journal of Agricultural and Environmental Ethics* 31, no. 4 (2018): 483–498.

Buchthal, Joanna, Sam Weiss Evans, Jeantine Lunshof, Sam R. Telford III, and Kevin M. Esvelt. "Mice Against Ticks: An Experimental Community-guided Effort to Prevent Tick-borne Disease by Altering the Shared Environment." *Philosophical Transactions of the Royal Society B* 374, no. 1772 (2019): 20180105.

Burg, T. M., and J. P. Croxall. "Global Population Structure and Taxonomy of the Wandering Albatross Species Complex." *Molecular Ecology* 13, no. 8 (2004): 2345–2355.

Callen, Alex, Matt W. Hayward, Kaya Klop-Toker, Benjamin L. Allen, Guy Ballard, Chad T. Beranek, Femke Broekhuis, et al. "Envisioning the Future with 'Compassionate Conservation': An Ominous Projection for Native Wildlife and Biodiversity." *Biological Conservation* 241 (2020): 108365.

Campbell-Smith, Jennifer. "A Viral Coyote-Badger Video Demonstrates the Incredible Complexity of Nature." *High Country News.* February 14, 2020. https://www.hcn.org /issues/52.4/wildlife-a-viral-coyote-badger-video-demonstrates-the-incredible -complexity-of-nature.

Carson, Rachel. *Silent Spring.* New York: Houghton Mifflin Harcourt, 2002.

Cäsar, Cristiane, Klaus Zuberbühler, Robert J. Young, and Richard W. Byrne. "Titi Monkey Call Sequences Vary with Predator Location and Type." *Biology Letters* 9, no. 5 (2013): 20130535.

Cassani Davis, Lauren. "Do Emotions and Morality Mix?" *The Atlantic.* February 5, 2016. https://www.theatlantic.com/science/archive/2016/02/how-do-emotions-sway-moral -thinking/460014/.

Cassel, Katie. *Nā Pua o Kōkeʻe: Field Guide to the Native Flowering Plants of Northwestern Kauaʻi.* Rancho Palos Verdes, CA: Quaking Aspen Books, 2010.

Chisholm, Suzanne and Michael Parfit, dirs. *Saving Luna.* 2007. [film]

Chiyo, Patrick I., Vincent Obanda, and David K. Korir. "Illegal Tusk Harvest and the Decline of Tusk Size in the African Elephant." *Ecology and Evolution* 5, no. 22 (2015): 5216–5229.

Clark, Stephen R. L. "Animals in Classical and Late Antique Philosophy." In *The Oxford Handbook of Animal Ethics*, edited by Tom L. Beauchamp and R. G. Frey. New York: Oxford University Press, 2011.

Clarkson, Chris, Zenobia Jacobs, Ben Marwick, Richard Fullagar, Lynley Wallis, Mike Smith, Richard G. Roberts, et al. "Human Occupation of Northern Australia by 65,000 Years Ago." *Nature* 547, no. 7663 (2017): 306–310.

Clayton, Susan, John Fraser, and Carol D. Saunders. "Zoo Experiences: Conversations, Connections, and Concern for Animals." Special issue, published in affiliation with the American Zoo and Aquarium Association, *Zoo Biology* 28, no. 5 (2009): 377–397.

Clayton, Susan, John Fraser, and Claire Burgess. "The Role of Zoos in Fostering Environmental Identity." *Ecopsychology* 3, no. 2 (2011): 87–96.

Clément, Marion Alice. "Habitat Features and Behavioral Plasticity Promote Barred Owl Presence in Developed Landscapes." Master's thesis, Clemson University, 2020. https://tigerprints.clemson.edu/all_theses/3364.

Cohen, Jon. "Commission Charts Narrow Path for Editing Human Embryos." *Science.* September 3, 2020. https://www.sciencemag.org/news/2020/09/commission-charts-narrow-path-editing-human-embryos.

Companion Animals New Zealand. Companion Animals in New Zealand 2020. Auckland, New Zealand. Accessed at https://static1.squarespace.com/static/5d1bf13a3f8e88000 1289eeb/t/5f768e8a17377653bd1eebef/1601605338749/Companion+Animals+in+NZ +2020+%281%29.pdf.

Connors, Martha Schindler. "Do Wolfdogs Make Good Pets?" *The Bark.* November/December 2010.

Cook, Robert G., and Catherine Fowler. "'Insight' in Pigeons: Absence of Means–end Processing in Displacement Tests." *Animal Cognition* 17, no. 2 (2014): 207–220.

Cottingham, John. "'A Brute to the Brutes?': Descartes' Treatment of Animals." *Philosophy* 53, no. 206 (1978): 551–559.

Cowperthwaite, Gabriela, dir. *Blackfish.* 2013. [film]

Cubaynes, Sarah, Daniel R. MacNulty, Daniel R. Stahler, Kira A. Quimby, Douglas W. Smith, and Tim Coulson. "Density-dependent Intraspecific Aggression Regulates Survival in Northern Yellowstone Wolves (*Canis lupus*)." *Journal of Animal Ecology* 83, no. 6 (2014): 1344–1356.

Cuthbert, Richard J., John Cooper, and Peter G. Ryan. "Population Trends and Breeding Success of Albatrosses and Giant Petrels at Gough Island in the Face of At-sea and On-land Threats." *Antarctic Science* 26, no. 2 (2014): 163.

Daley, Jason. "North America's Rarest Warbler Comes Off the Endangered List." *Smithsonian*. October 11, 2019. https://www.smithsonianmag.com/smart-news/north -americas-rarest-warbler-comes-endangered-list-180973324/.

Darimont, Chris T., Stephanie M. Carlson, Michael T. Kinnison, Paul C. Paquet, Thomas E. Reimchen, and Christopher C. Wilmers. "Human Predators Outpace Other Agents of Trait Change in the Wild." *Proceedings of the National Academy of Sciences* 106, no. 3 (2009): 952–954.

Darwin, Charles. *The Descent of Man, and Selection in Relation to Sex.* United Kingdom: D. Appleton, 1872.

Das, Ushnik, Anshu Kumari, Shruthi Sharma, and Laxmi T. Rao. "Demonstration of Altruistic Behaviour in Rats." *bioRxiv* (2019): 805481.

Da Silva, Maria N. F., Shepard H. Shepard Jr, and Douglas W. Yu. "Conservation Implications of Primate Hunting Practices Among the Matsigenka of Manu National Park." *Neotropical Primates* 13.2 (2005): 31–36.

Delaney, Ted. "Ota Benga (d. 1916)" *Encyclopedia Virginia*. October 23, 2020. Retrieved from http://www.EncyclopediaVirginia.org/Benga_Ota_ca_1883-1916.

Department of Conservation (NZ). "History of Kapiti Island." https://www.doc.govt.nz/parks -and-recreation/places-to-go/wellington-kapiti/places/kapiti-island-nature-reserve /historic-kapiti-island/.

Department of Conservation (NZ). "Kapiti Island Nature and Conservation." https://www .doc.govt.nz/parks-and-recreation/places-to-go/wellington-kapiti/places/kapiti-island -nature-reserve/nature-and-conservation/.

Department of the Environment and Energy (Australia). *Threatened Species Strategy— Year Three Progress Report.* 2019.

De Waal, Frans. *Are We Smart Enough to Know How Smart Animals Are?* New York: W. W. Norton & Company, 2016.

DIISE. The Database of Island Invasive Species Eradications, developed by Island Conservation, Coastal Conservation Action Laboratory UCSC, IUCN SSC Invasive

Species Specialist Group, University of Auckland and Landcare Research New Zealand. http://diise.islandconservation.org. 2018.

Doherty, Tim S., Alistair S. Glen, Dale G. Nimmo, Euan G. Ritchie, and Chris R. Dickman. "Invasive Predators and Global Biodiversity Loss." *Proceedings of the National Academy of Sciences* 113, no. 40 (2016): 11261–11265.

Doherty, Tim S., Don A. Driscoll, Dale G. Nimmo, Euan G. Ritchie, and Ricky-John Spencer. "Conservation or Politics? Australia's Target to Kill 2 Million Cats." *Conservation Letters* 12, no. 4 (2019): e12633.

Donaldson, Sue and Will Kymlicka. *Zoopolis: A Political Theory of Animal Rights.* New York: Oxford University Press, 2011.

Doughton, Sandi. "Elephant Chai Suffered Injuries, Weight Loss Months before Her Death." *Seattle Times.* March 19, 2016. https://www.seattletimes.com/seattle-news/times-watchdog/chai-suffered-injuries-weight-loss-months-before-her-death-in-oklahoma/.

Dowie, Mark. *Conservation Refugees: The Hundred-year Conflict between Global Conservation and Native Peoples.* Cambridge, MA: MIT Press, 2009.

Driver, George. "Fact check: Has NZ Spent More on Corrections in Two Years than on All Treaty of Waitangi Settlements?" Radio New Zealand. October 8, 2020. https://www.rnz.co.nz/news/political/427913/fact-check-has-nz-spent-more-on-corrections-in-two-years-than-on-all-treaty-of-waitangi-settlements.

Eckert, Rainer. "On the Cult of the Snake in Ancient Baltic and Slavic Tradition (Based on Language Material from the Latvian Folksongs)." *Zeitschrift für Slawistik* 43, no. 1 (1998): 94–100.

"Elsa's Legacy: The Born Free Story" *Nature.* PBS. Season 29 Episode 6. Aired: 01/08/2011. https://www.pbs.org/video/nature-elsas-legacy-the-born-free-story/.

Esvelt, Kevin M., Andrea L. Smidler, Flaminia Catteruccia, and George M. Church. "Emerging Technology: Concerning RNA-guided Gene Drives for the Alteration of Wild Populations." *Elife* 3 (2014): e03401.

Esvelt, Kevin M., and Neil J. Gemmell. "Conservation Demands Safe Gene Drive." *PLoS Biology* 15, no. 11 (2017): e2003850.

Evans, Scott D., Ian V. Hughes, James G. Gehling, and Mary L. Droser. "Discovery of the Oldest Bilaterian from the Ediacaran of South Australia." *Proceedings of the National Academy of Sciences* 117, no. 14 (2020): 7845–7850.

Farnsworth, John. "The Condor Question Revisited." *Minding Nature* 8, no. 2 (Spring 2015).

Faurby, Søren, Daniele Silvestro, Lars Werdelin, and Alexandre Antonelli. "Brain Expansion in Early Hominins Predicts Carnivore Extinctions in East Africa." *Ecology Letters* 23, no. 3 (2020): 537–544.

Fickett-Wilbar, David. "Cernunnos: Looking a Different Way." *Proceedings of the Harvard Celtic Colloquium* 23 (2003): 80–111.

Finley, William L. "Life History of the California Condor Part IV—The Young Condor in Captivity." *The Condor* 12, no. 1 (1910): 4–11.

Francis of Assisi, *The Little Flowers of Saint Francis of Assisi*. ed. and trans. Roger Hudleston. Italy: Limited Editions Club, 1930.

Fraser, John Robert. "An Examination of Environmental Collective Identity Development across Three Life-Stages: The Contribution of Social Public Experiences at Zoos." PhD diss., Antioch University, 2009.

Freeman, Benjamin G., Micah N. Scholer, Viviana Ruiz-Gutierrez, and John W. Fitzpatrick. "Climate Change Causes Upslope Shifts and Mountaintop Extirpations in a Tropical Bird Community." *Proceedings of the National Academy of Sciences* 115, no. 47 (2018): 11982–11987.

Freeman, Mark. "Return of the Sacred Condor," *Medford Mail Tribune*. April 4, 2019. https://mailtribune.com/news/top-stories/condors-planned-for-release-in-northern-california-could-show-up-in-oregon.

Fitzpatrick, Benjamin M., and H. Bradley Shaffer. "Hybrid Vigor Between Native and Introduced Salamanders Raises New Challenges for Conservation." *Proceedings of the National Academy of Sciences* 104, no. 40 (2007): 15793–15798.

Food and Agriculture Organization of the United Nations. "Land Use in Agriculture by the Numbers." May 7, 2020. http://www.fao.org/sustainability/news/detail/en/c/1274219/.

Foot, Philippa. "The Problem of Abortion and the Doctrine of Double Effect." *Oxford Review* 5 (1967): 5–15.

Foote, Andrew D., Nagarjun Vijay, María C. Ávila-Arcos, Robin W. Baird, John W. Durban, Matteo Fumagalli, Richard A. Gibbs, et al. "Genome-culture Coevolution Promotes Rapid Divergence of Killer Whale Ecotypes." *Nature Communications* 7, no. 1 (2016): 1–12.

Forest & Bird. "Frequently Asked Questions about 1080." April 16, 2018. https://www
.forestandbird.org.nz/resources/frequently-asked-questions-about-1080.

Foster, John W. "Wings of the Spirit: The Place of the California Condor Among Native Peoples
of the Californias." Accessed December 1, 2020. California Department of Parks and
Recreation. https://www.parks.ca.gov/?page_id=23527.

Funk, Cary, Meg Hefferon. "Public Views of Gene Editing for Babies Depend on How It
Would Be Used." Pew Research Center, July 26, 2018. https://www.pewresearch.org
/science/2018/07/26/public-views-of-gene-editing-for-babies-depend-on-how-it
-would-be-used/.

Fuouco, Linda Wilson, Chico Harlan. "Wolf Dogs Killed Owner, Autopsy Determines."
Pittsburgh Post-Gazette, July 18, 2006. https://www.post-gazette.com/local/west
moreland/2006/07/19/Wolf-dogs-killed-owner-autopsy-determines/stories/2006
07190197.

Galetti, Mauro, Marcos Moleón, Pedro Jordano, Mathias M. Pires, Paulo R. Guimaraes Jr,
Thomas Pape, Elizabeth Nichols, et al. "Ecological and Evolutionary Legacy of Megafauna
Extinctions." *Biological Reviews* 93, no. 2 (2018): 845–862.

George, Andrew, ed. *The Epic of Gilgamesh: The Babylonian Epic Poem and Other Texts in
Akkadian and Sumerian*. New York: Penguin, 2002.

Gheorghe, Adrian. "The Grass Snake." Last modified January 24, 2011. http://alexisphoenix
.org/romaniasnake.php.

Gibbens, Sarah. "Heart-Wrenching Video Shows Starving Polar Bear on Iceless Land." *National
Geographic*. December 7, 2017. https://www.nationalgeographic.com/news/2017/12/polar
-bear-starving-arctic-sea-ice-melt-climate-change-spd/.

Gibbons, Ann. "Meet the Frail, Small-brained People who First Trekked out of Africa."
Science, November 22, 2016. doi:10.1126/ science.aal0416.

Gilbert, Nathalie. "Movement and Foraging Ecology of Partially Migrant Birds in a Changing
World." Doctoral thesis, University of East Anglia, 2015. https://ueaeprints.uea.ac.uk
/id/eprint/59626.

Gill, Jacquelyn L., John W. Williams, Stephen T. Jackson, Katherine B. Lininger, and Guy S.
Robinson. "Pleistocene Megafaunal Collapse, Novel Plant Communities, and Enhanced
Fire Regimes in North America." *Science* 326, no. 5956 (2009): 1100–1103.

Greco, Brian J., Cheryl L. Meehan, Jen N. Hogan, Katherine A. Leighty, Jill Mellen, Georgia J. Mason, and Joy A. Mench. "The Days and Nights of Zoo Elephants: Using Epidemiology to Better Understand Stereotypic Behavior of African Elephants (*Loxodonta africana*) and Asian Elephants (*Elephas maximus*) in North American Zoos." *PLoS One* 11, no. 7 (2016).

Greene, Joshua D., R. Brian Sommerville, Leigh E. Nystrom, John M. Darley, and Jonathan D. Cohen. "An fMRI Investigation of Emotional Engagement in Moral Judgment." *Science* 293, no. 5537 (2001): 2105–2108.

Gruen, Lori. *Entangled Empathy: An Alternative Ethic for Our Relationships with Animals.* Brooklyn: Lantern Books, 2015.

Guernsey, Brenda. "Constructing the Wilderness and Clearing the Landscape: A Legacy of Colonialism in Northern British Columbia." *Landscapes of Clearance: Archaeological and Anthropological Perspectives* 57 (2008): 112.

Guynup, Sharon. "Captive Tigers in the U.S. Outnumber Those in the Wild. It's a Problem." *National Geographic.* December 2019. https://www.nationalgeographic.com/animals /2019/11/tigers-in-the-united-states-outnumber-those-in-the-wild-feature/.

Haig, Susan M., Thomas D. Mullins, Eric D. Forsman, Pepper W. Trail, and L. I. V. Wennerberg. "Genetic Identification of Spotted Owls, Barred Owls, and Their Hybrids: Legal Implications of Hybrid Identity." *Conservation Biology* 18, no. 5 (2004): 1347–1357.

Hancocks, David. *A Different Nature: The Paradoxical World of Zoos and Their Uncertain Future.* Berkeley and Los Angeles: University of California Press, 2001.

Hancocks, David. "Bamboo Should Be Sent to a Place Where She Can Heal." *Seattle Times.* September 9, 2005. https://www.seattletimes.com/opinion/bamboo-should-be-sent-to -a-place-where-she-can-heal/.

Hancocks, David. "Hansa's Short Life One of Deprivation." *Seattle PI.* June 11, 2007. Updated: March 21, 2011. https://www.seattlepi.com/local/opinion/article/Hansa-s-short-life -one-of-deprivation-1240222.php.

Hancocks, David. "Former Woodland Park Head Sees Sad Future for Elephants in OK." Crosscut. April 2, 2015. https://crosscut.com/2015/04/guest-opinion-former-woodland -park-head-sees-sad-future-for-elephants-in-ok.

Harris, David R., and Gordon C. Hillman, eds. *Foraging and Farming: The Evolution of Plant Exploitation.* United Kingdom: Routledge, 2014.

Hart, Donna. *Man the Hunted: Primates, Predators, and Human Evolution*. United Kingdom: Routledge, 2018.

Hart, George. *The Routledge Dictionary of Egyptian Gods and Goddesses*. United Kingdom: Routledge, 2005.

Hayward, Matt W., Alex Callen, Benjamin L. Allen, Guy Ballard, Femke Broekhuis, Cassandra Bugir, Rohan H. Clarke, et al. "Deconstructing Compassionate Conservation." *Conservation Biology* 33, no. 4 (2019): 760–768.

Hebblewhite, Mark, and Jesse Whittington. "Wolves Without Borders: Transboundary Survival of Wolves in Banff National Park over Three Decades." *Global Ecology and Conservation* (2020): e01293.

Hecht, Susanna B. *The Scramble for the Amazon and the "Lost Paradise" of Euclides Da Cunha*. Chicago: University of Chicago Press, 2013.

Hedrick, Ann V. "The Development of Animal Personality." *Frontiers in Ecology and Evolution* 5 (2017): 14.

Hennessy, Elizabeth. *On the Backs of Tortoises: Darwin, the Galápagos, and the Fate of an Evolutionary Eden*. New Haven: Yale University Press, 2019.

Hess, Elizabeth. *Nim Chimpsky: The Chimp Who Would Be Human*. New York: Bantam, 2008.

Higgs, Eric. *Nature by Design: People, Natural Process, and Ecological Restoration*. Cambridge, MA: MIT Press, 2003.

Hirata, Satoshi, Kunio Watanabe, and Kawai Masao. "'Sweet-Potato Washing' Revisited." In *Primate Origins of Human Cognition and Behavior*, 487–508. Tokyo: Springer, 2008.

Hirschman, Elizabeth C. "Consumers and Their Animal Companions." *Journal of Consumer Research* 20, no. 4 (1994).

Hitchmough, R., van Winkel, D., Lettink, M. & Chapple, D. 2019. Hoplodactylus delcourti. The IUCN Red List of Threatened Species 2019: e.T10254A120158840. https://dx.doi.org /10.2305/IUCN.UK.2019-2.RLTS.T10254A120158840.en. Downloaded on October 21, 2020.

Holmes, Nick D., Dena R. Spatz, Steffen Oppel, Bernie Tershy, Donald A. Croll, Brad Keitt, Piero Genovesi, et al. "Globally Important Islands Where Eradicating Invasive Mammals Will Benefit Highly Threatened Vertebrates." *PloS One* 14, no. 3 (2019).

Holsman, Melissa. "New Details Emerge About Elephant Deaths at Fellsmere Center." *TCPalm*. February 5, 2017. https://www.tcpalm.com/story/news/local/indian-river

-county/2017/04/05/new-details-emerge-elephant-deaths-fellsmere-center
/99748330/.

Hopper, Tristan. "Everyone Was Dead: When Europeans First Came to B.C., They Stepped Into the Aftermath of a Holocaust." *The National Post*. February 21, 2017. https://nationalpost.com/news/canada/everyone-was-dead-when-europeans-first-came-to-b-c-they-confronted-the-aftermath-of-a-holocaust.

Horta, Oscar. "Animal Suffering in Nature: The Case for Intervention." *Environmental Ethics* 39, no. 3 (2017): 261–279.

Hribal, Jason. *Fear of the Animal Planet: The Hidden History of Animal Resistance*. Chico, CA: AK Press, 2011.

Hunt, Heather, Warne, Kennedy. *It's My Egg (and You Can't Have It!)*. New Zealand: Potton & Burton, 2017.

Hunt, Terry L., and Carl P. Lipo. "The Last Great Migration: Human Colonization of the Remote Pacific Islands." Chapter 8 in *Human Dispersal and Species Movement: From Prehistory to the Present*, edited by Nicole Boivin, Rémy Crassard, and Michael Petraglia, 194–216. Cambridge: Cambridge University Press, 2017. doi:10.1017/9781316686942.009.

Hursthouse, Rosalind. "Virtue Ethics and the Treatment of Animals." In *The Oxford Handbook of Animal Ethics*, edited by Tom L. Beauchamp and R. G. Frey, 119–143. New York: Oxford University Press, 2011.

Isabella, Jude. "From Prejudice to Pride," *Hakai Magazine*, October 10, 2017. https://www.hakaimagazine.com/features/prejudice-pride/.

Isden, Jess, Carmen Panayi, Caroline Dingle, and Joah Madden. "Performance in Cognitive and Problem-Solving Tasks in Male Spotted Bowerbirds Does Not Correlate with Wating Success." *Animal Behaviour* 86, no. 4 (2013): 829–838.

Jackson, Stephen T. "Perspective: Ecological Novelty Is Not New." In *Novel Ecosystems: Intervening in the New Ecological World Order*, edited by R. J. Hobbs, E. S. Higgs and C. M. Hal, 63–65. Hoboken, NJ: John Wiley & Sons, 2013.

Jacobson, Sarah L., Stephen R. Ross, and Mollie A. Bloomsmith. "Characterizing Abnormal Behavior in a Large Population of Zoo-housed Chimpanzees: Prevalence and Potential Influencing Factors." *PeerJ* 4 (2016): e2225.

Jervis, Lori L., Paul Spicer, William C. Foster, Jeffrey Kelly, and Eli Bridge. "Resisting Extinction." *Conservation & Society* 17, no. 3 (2019): 227–235.

Johnson, Christopher N. "Ecological Consequences of Late Quaternary Extinctions of Megafauna." *Proceedings of the Royal Society B: Biological Sciences* 276, no. 1667 (2009): 2509–2519.

Johnson, Christopher N., Susan Rule, Simon G. Haberle, A. Peter Kershaw, G. Merna McKenzie, and Barry W. Brook. "Geographic Variation in the Ecological Effects of Extinction of Australia's Pleistocene Megafauna." *Ecography* 39, no. 2 (2016): 109–116.

Joint Secretariat. *Inuvialuit and Nanuq: A Polar Bear Traditional Knowledge Study*. Joint Secretariat, Inuvialuit Settlement Region. 2015.

Jørgensen, Dolly. "Conservation Implications of Parasite Co-reintroduction." *Conservation Biology* 29, no. 2 (2015): 602–604.

Kane, Will. "California Wolf Is Back in Oregon." SFGate. March 3, 2012, Updated December 24, 2013. https://www.sfgate.com/science/article/California-wolf-is-back-in -Oregon-3377534.php.

Kawall, Jason. "A History of Environmental Ethics," in *The Oxford Handbook of Environmental Ethics*, edited by Gardiner, Stephen Mark, and Allen Thompson. New York: Oxford University Press, 2017.

Keeling, Linda J., Jeff Rushen, and Ian J. H, Duncan. "Understanding Animal Welfare." In *Animal Welfare*, ed. Michael C. Appleby (Wallingford, Oxfordshire, UK & Cambridge, MA: CABI, 2011), 19–20.

Kelleher, Shannon R., Aimee J. Silla, and Phillip G. Byrne. "Animal Personality and Behavioral Syndromes in Amphibians: A Review of the Evidence, Experimental Approaches, and Implications for Conservation." *Behavioral Ecology and Sociobiology* 72, no. 5 (2018): 79.

Kimmerer, Robin Wall. *Braiding Sweetgrass: Indigenous Wisdom, Scientific Knowledge and the Teachings of Plants*. Minneapolis: Milkweed Editions, 2013.

King, Carolyn M. *The Handbook of New Zealand Mammals*, 2nd ed. New York: Oxford University Press.

Koch, Alexander, Chris Brierley, Mark M. Maslin, and Simon L. Lewis. "Earth System Impacts of the European Arrival and Great Dying in the Americas after 1492." *Quaternary Science Reviews* 207 (2019): 13–36.

Kofler, Natalie, James P. Collins, Jennifer Kuzma, Emma Marris, Kevin Esvelt, Michael Paul Nelson, Andrew Newhouse, et al. "Editing Nature: Local Roots of Global Governance." *Science* 362, no. 6414 (2018): 527–529.

Kohda, Masanori, Hatta Takashi, Tmohiro Takeyama, Satoshi Awata, Hirokazu Tanaka, Jun-ya Asai, and Alex Jordan. "Cleaner Wrasse Pass the Mark Test. What Are the Implications for Consciousness and Self-awareness Testing in Animals?." *bioRxiv* (2018): 397067.

Laidman, Jenni. "Zoos Using Drugs to Help Manage Anxious Animals." *The Blade.* September 12, 2005. https://www.toledoblade.com/frontpage/2005/09/12/Zoos-using-drugs-to-help-manage-anxious-animals.html.

Landers, Rich. "Wolf 47 Works Full-Time for Washington Wildlife Researchers." *The Spokesman-Review.* March 9, 2014. https://www.spokesman.com/stories/2014/mar/09/wolf-47-works-full-time-for-washington-wildlife/.

LaPointe, Dennis A., Carter T. Atkinson, and Michael D. Samuel. "Ecology and Conservation Biology of Avian Malaria." *Annals of the New York Academy of Sciences* 1249, no. 1 (2012): 211–226.

Lazzaroni, Martina, Friederike Range, Jessica Backes, Katrin Portele, Katharina Scheck, and Sarah Marshall-Pescini. "The Effect of Domestication and Experience on the Social Interaction of Dogs and Wolves with a Human Companion." *Frontiers in Psychology* 11 (2020): 785.

Le Mitouard, Eric. "Sa pétition pour sauver les rats de Paris a déjà recueilli 17,000 signatures." *Le Parisien.* December 14, 2016. https://www.leparisien.fr/paris-75/paris-75005/sa-petition-pour-sauver-les-rats-de-paris-a-deja-recueilli-17-000-signatures-14-12-2016-6456072.php.

Lenders, H. J., and Ingo AW Janssen. "The Grass Snake and the Basilisk: From Pre-Christian Protective House God to the Antichrist." *Environment and History* 20, no. 3 (2014): 319–346.

Leopold, Aldo. *A Sand County Almanac, and Sketches Here and There.* New York: Oxford University Press, 1989.

Lescureux, Nicolas and John D. C. Linnell. "Warring Brothers: The Complex Interactions between Wolves (*Canis lupus*) and Dogs (*Canis familiaris*) in a Conservation Context." *Biological Conservation* 171 (2014): 232–245. ScienceDirect.

Lewis, Dyani. "An Identity Crisis for the Australian Dingo." *UnDark.* August 12, 2019. https://undark.org/2019/08/12/identity-crisis-australian-dingo/.

Li, Fay-Wei, Juan Carlos Villarreal, Steven Kelly, Carl J. Rothfels, Michael Melkonian, Eftychios Frangedakis, Markus Ruhsam, et al. "Horizontal Transfer of an Adaptive Chimeric Photoreceptor from Bryophytes to Ferns." *Proceedings of the National Academy of Sciences* 111, no. 18 (2014): 6672–6677.

Linklater, Wayne, and Jamie Steer. "Predator Free 2050: A Flawed Conservation Policy Displaces Higher Priorities and Better, Evidence-based Alternatives." *Conservation Letters* 11, no. 6 (2018): e12593.

Linn, A., K. J. Burns, and C. H. Richart. "Red-crested Cardinal (*Paroaria coronata*)," *Cornell Lab of Ornithology: Birds of the World*, March 4, 2020. https://doi.org/10.2173/bow .reccar.01.

Littin, K. E., C. E. O'Connor, and C. T. Eason. "Comparative Effects of Brodifacoum on Rats and Possums." *New Zealand Plant Protection* 53 (2000): 310–315.

Littin, K., P. Fisher, N. J. Beausoleil, and T. Sharp. "Welfare Aspects of Vertebrate Pest Control and Culling: Ranking Control Techniques for Humaneness." *Revue Scientifique et Technique (International Office of Epizootics)* 33, no. 1 (2014): 281–89.

Loh, Tse-Lynn, Eric R. Larson, Solomon R. David, Lesley S. de Souza, Rebecca Gericke, Mary Gryzbek, Andrew S. Kough, Philip W. Willink, and Charles R. Knapp. "Quantifying the Contribution of Zoos and Aquariums to Peer-reviewed Scientific Research." *Facets* 3, no. 1 (2018): 287–299.

Looper, Matthew G. *The Beast Between: Deer Imagery in Ancient Maya Art.* Austin: University of Texas Press, 2019.

Loss, Scott R., Tom Will, and Peter P. Marra. "The Impact of Free-ranging Domestic Cats on Wildlife of the United States." *Nature Communications* 4, no. 1 (2013): 1–8.

Low, Philip, Jaak Panksepp, Diana Reiss, David Edelman, Bruno Van Swinderen, and Christof Koch. "The Cambridge Declaration on Consciousness." Presented at In Francis Crick Memorial Conference, Cambridge, England, 2012.

Lundgren, Erick J., Daniel Ramp, William J. Ripple, and Arian D. Wallach. "Introduced Megafauna Are Rewilding the Anthropocene." *Ecography* 41, no. 6 (2018): 857–866.

MacDonald, Edith A., Jovana Balanovic, Eric D. Edwards, Wokje Abrahamse, Bob Frame, Alison Greenaway, Robyn Kannemeyer, et al. "Public Opinion Towards Gene Drive As a

Pest Control Approach for Biodiversity Conservation and the Association of Underlying Worldviews." *Environmental Communication* (2020): 1–15.

Madrid, Cienna. "Cash Cows." *The Stranger.* May 11, 2011. https://www.thestranger.com /seattle/cash-cows/Content?oid=8078780.

Mallet, James, Nora Besansky, and Matthew W. Hahn. "How Reticulated Are Species?" *BioEssays* 38, no. 2 (2016): 140–149.

Mapes, Lynda. "How Quickly They Grow: Hansa Is 5." *Seattle Times.* November 6, 2005. https://www.seattletimes.com/seattle-news/how-quickly-they-grow-hansa-is-5/.

Mapes, Lynda. "Seattle Zoo's Beloved Young Elephant Dies." *Seattle Times.* June 9, 2007. https://www.seattletimes.com/seattle-news/seattle-zoos-beloved-young-elephant -dies/.

Marino, Lori, Naomi A. Rose, Ingrid N. Visser, Heather Rally, Hope Ferdowsian, and Veronica Slootsky. "The Harmful Effects of Captivity and Chronic Stress on the Well-being of Orcas (*Orcinus orca*)." *Journal of Veterinary Behavior* 35 (2020): 69–82.

Marris, Emma. *Rambunctious Garden: Saving Nature in a Post-wild World.* New York: Bloomsbury Publishing, 2013.

Marris, Emma. "Wolf Cull Will Not Save Threatened Canadian Caribou." *Nature,* January 20, 2015. https://www.nature.com/news/wolf-cull-will-not-save-threatened -canadian-caribou-1.16734.

Marris, Emma. "Why OR7 Is a Celebrity." *High Country News.* January 23, 2017. https://www .hcn.org/issues/49.1/why-or7-is-a-celebrity.

Marris, Emma. "Resurrecting a Long-Lost Galapagos Giant Tortoise." *Wired.* September 14, 2017. https://www.wired.com/story/resurrecting-a-long-lost-galapagos-giant-tortoise/.

Marris, Emma. "Process of Elimination." *Wired.* February 28, 2018. https://www.wired.com /story/crispr-eradicate-invasive-species/.

Marris, Emma. "Large Island Declared Rat-Free in Biggest Removal Success." *National Geographic.* May 9, 2018. https://www.nationalgeographic.com/news/2018/05/south -georgia-island-rat-free-animals-spd/.

Marris, Emma. "These Rare Zebras Are Dependent on Humans, for Now." *National Geographic.* January 16, 2020. https://www.nationalgeographic.com/animals/2020/01 /grevys-zebras-face-drought-hay-deliveries/.

Mason, Georgia J., and N. Latham. "Can't Stop, Won't Stop: Is Stereotypy a Reliable Animal Welfare Indicator?" *Animal Welfare* 13, Suppl. 1 (2004): 57–69.

Masserman, Jules H., Stanley Wechkin, and William Terris. "'Altruistic' Behavior in Rhesus Monkeys." *American Journal of Psychiatry* 121, no. 6 (1964): 584–585.

Mathews, Freya. "Wild Animals Are Starving, and It's Our Fault, so Should We Feed Them?" *The Conversation*, August 18, 2013. https://theconversation.com/wild-animals -are-starving-and-its-our-fault-so-should-we-feed-them-16803.

McGraw, James and William Foster. "When You Stand Up for Zoos You Stand Up for Elephants." *Seattle Times*, June 15, 2007. https://www.seattletimes.com/opinion/when -you-stand-up-for-zoos-you-stand-up-for-elephants/.

McMahan, Jeff. "The Meat Eaters." *New York Times*. September 19, 2010. https://opinionator .blogs.nytimes.com/2010/09/19/the-meat-eaters/.

Meany, Edmond Stephen. *Vancouver's Discovery of Puget Sound: Portraits and Biographies of the Men Honored in the Naming of Geographic Features of Northwestern America.* New York: Macmillan, 1915.

Mech, L. David, and Luigi Boitani, eds. *Wolves: Behavior, Ecology, and Conservation.* Chicago: University of Chicago Press, 2010.

Meretsky, Vicky J., Noel FR Snyder, Steven R. Beissinger, David A. Clendenen, and James W. Wiley. "Demography of the California Condor: Implications for Reestablishment." *Conservation Biology* 14, no. 4 (2000): 958.

Middle Island Project. *2017–2018 Penguin Breeding Season—Completion Report.* http://www .warrnamboolpenguins.com.au/sites/warrnamboolpenguins.com.au/files/documents /2017-18%20Completion%20Report.pdf.

Ministry for Culture and Heritage (NZ). "The Ngāi Tahu Claim." New Zealand History. Updated June 14, 2016. https://nzhistory.govt.nz/politics/treaty/the-treaty-in-practice /ngai-tahu.

Ministry for Culture and Heritage (NZ). "The Treaty in Brief." New Zealand History. Updated May 17, 2017. https://nzhistory.govt.nz/politics/treaty/the-treaty-in-brief.

Minta, Steven C., Kathryn A. Minta, and Dale F. Lott. "Hunting Associations Between Badgers (*Taxidea taxus*) and Coyotes (*Canis latrans*)." *Journal of Mammalogy* 73, no. 4 (1992): 814–820.

Miskelly, Colin. "Extinct Birds of New Zealand—A Diverse Menagerie, Sadly Departed." Te Papa Blog. July 28, 2015. https://blog.tepapa.govt.nz/2015/07/28/extinct-birds -of-new-zealand-part-1-a-diverse-menagerie-sadly-departed/.

Mittermeier, Cristina. "Starving-Polar-Bear Photographer Recalls What Went Wrong." *National Geographic*. August 2018. https://www.nationalgeographic.com/magazine/2018 /08/explore-through-the-lens-starving-polar-bear-photo/.

Mock, Jillian. "Lead Ammo, the Top Threat to Condors, Is Now Outlawed in California." *Audubon*. July 1, 2019. https://www.audubon.org/news/lead-ammo-top-threat-condors -now-outlawed-california.

Mollman, Steve. "Attendance at SeaWorld San Diego Has Plummeted since the 'Blackfish' Documentary." *Quartz*. May 26, 2016. https://qz.com/692900/attendance-at-seaworld -san-diego-has-plummeted-since-the-blackfish-documentary/.

Montague, Michael J., Gang Li, Barbara Gandolfi, Razib Khan, Bronwen L. Aken, Steven M. J. Searle, Patrick Minx, et al. "Comparative Analysis of the Domestic Cat Genome Reveals Genetic Signatures Underlying Feline Biology and Domestication." *Proceedings of the National Academy of Sciences* 111, no. 48 (2014): 17230–17235.

Morell, Virginia. *Animal Wise: The Thoughts and Emotions of Our Fellow Creatures*. New York: Crown Publishers, 2013.

Morell, Virginia. "Snakes have friends too." *National Geographic*, May 13, 2020, https://www.nationalgeographic.com/animals/2020/05/snakes-have-friends-adding-to -evidence-animal-sociability/.

Mulady, Kathy. "Donation Made to Honor Hansa." *Seattle PI*, June 11, 2007. Updated March 21, 2011. https://www.seattlepi.com/local/article/Donation-made-to-honor -Hansa-1240282.php.

Nagel, Thomas. "What Is It Like to Be a Bat?" *The Philosophical Review* 83, no. 4 (1974): 435–450.

Nadasdy, Paul. *Sovereignty's Entailments: First Nation State Formation in the Yukon*. Toronto: University of Toronto Press, 2017.

National Academy of Sciences 2020. *Heritable Human Genome Editing*. Washington, DC: The National Academies Press, 2020.

National Audubon Soc. v. Hester, 627 F. Supp. 1419 (D.D.C. 1986) U.S. District Court for the District of Columbia—627 F. Supp. 1419 (D.D.C. 1986) February 3, 1986.

Nengo, Isaiah, Paul Tafforeau, Christopher C. Gilbert, John G. Fleagle, Ellen R. Miller, Craig Feibel, David L. Fox, et al. "New Infant Cranium from the African Miocene Sheds Light on Ape Evolution." *Nature* 548, no. 7666 (2017): 169–174.

Newshub. "Baby Possums Drowned at Drury School's 'Inhumane' Fundraiser." February 7, 2017. https://www.newshub.co.nz/home/new-zealand/2017/07/baby-possums-drowned -at-drury-school-s-inhumane-fundraiser.html.

Nicholls, Henry. *The Galápagos: A Natural History.* New York: Basic Books, 2014.

Nicklaus, David. "Peabody CEO Earns $7.6 million but Will Be Hurt by Sinking Stock Price." *Saint Louis Post-Dispatch.* March 26, 2020. https://www.stltoday.com/business /columns/david-nicklaus/peabody-ceo-earns-7-6-million-but-will-be-hurt-by-sinking -stock-price/article_980a3987-bc31-50de-a1b9-70de0c47b28d.html.

Nietlisbach, Pirmin, Peter Wandeler, Patricia G. Parker, Peter R. Grant, B. Rosemary Grant, Lukas F. Keller, and Paquita EA Hoeck. "Hybrid Ancestry of an Island Subspecies of Galápagos Mockingbird Explains Discordant Gene Trees." *Molecular Phylogenetics and Evolution* 69, no. 3 (2013): 581–592.

Nihei, Yoshiaki, and Hiroyoshi Higuchi. "When and Where Did Crows Learn to Use Automobiles as Nutcrackers." *Tohoku Psychologica Folia* 60 (2001): 93–97.

Noble, Charleston, John Min, Jason Olejarz, Joanna Buchthal, Alejandro Chavez, Andrea L. Smidler, Erika A. DeBenedictis, George M. Church, Martin A. Nowak, and Kevin M. Esvelt. "Daisy-Chain Gene Drives for the Alteration of Local Populations." *Proceedings of the National Academy of Sciences* 116, no. 17 (2019): 8275–8282.

Norcross, Alastair. "Death for Animals." In *The Oxford Handbook of Philosophy of Death*, edited by Ben Bradley, Fred Feldman, and Jens Johansson. New York: Oxford University Press, 2013.

Norton, Bryan G., Michael Hutchins, Terry Maple, and Elizabeth Stevens, eds. *Ethics on the Ark: Zoos, Animal Welfare, and Wildlife Conservation.* Washington, DC: Smithsonian Institution, 1995.

Nussbaum, Martha C. *Frontiers of Justice: Disability, Nationality, Species Membership.* Boston: Harvard University Press, 2009.

Nuwer, Rachel. "The Strange and Dangerous World of America's Big Cat People." *Longreads*, March 2020. https://longreads.com/2020/03/16/tiger-trafficking-in-america/.

Nuwer, Rachel Love. *Poached: Inside the Dark World of Wildlife Trafficking.* United Kingdom: Hachette, 2018.

NZ Pocket Guide. "Why New Zealand Hates Possums." Accessed November 13, 2020. https://nzpocketguide.com/new-zealand-hates-possums/.

O'Leary, Maureen A., Jonathan I. Bloch, John J. Flynn, Timothy J. Gaudin, Andres Giallombardo, Norberto P. Giannini, Suzann L. Goldberg, et al. "The Placental Mammal Ancestor and the Post–K-Pg Radiation of Placentals." *Science* 339, no. 6120 (2013): 662–667.

Oregon Department of Fish and Wildlife. "Oregon Wolf Population." Accessed December 2, 2020. https://dfw.state.or.us/Wolves/population.asp.

Oregon Department of Fish and Wildlife. *Oregon Wolf Conservation and Management 2019 Annual Report*, 2020. https://www.dfw.state.or.us/Wolves/docs/oregon_wolf_program/2019_Annual_Wolf_Report_FINAL.pdf.

Oregon Wild. "Wolves Come Home to Oregon." Accessed December 2, 2020. https://oregonwild.org/wildlife/wolves-come-home-oregon.

Orlando, Ludovic. "Back to the Roots and Routes of Dromedary Domestication." *Proceedings of the National Academy of Sciences* 113, no. 24 (2016): 6588–6590.

Pagano, Anthony M., George M. Durner, Karyn D. Rode, Todd C. Atwood, Stephen N. Atkinson, Elizabeth Peacock, Daniel P. Costa, Megan A. Owen, and Terrie M. Williams. "High-energy, High-fat Lifestyle Challenges an Arctic Apex Predator, the Polar Bear." *Science* 359, no. 6375 (2018): 568–572.

Palmer, Clare. *Animal Ethics in Context*. New York: Columbia University Press, 2010.

Palmer, Clare. "Should We Provide the Bear Necessities? Climate Change, Polar Bears and the Ethics of Supplemental Feeding." In *Animals in Our Midst*, edited by Bernice Bovenkirk and Josef Keulartz. New York: Springer, forthcoming.

Parker, Ian. "Killing Animals at the Zoo." *New Yorker*. January 9, 2017. https://www.newyorker.com/magazine/2017/01/16/killing-animals-at-the-zoo.

Parrott, Marissa, Dale Nimmo, and Euan Ritchie. "How You Can Help—not Harm—Wild Animals Recovering from Bushfires." The Conversation. February 19, 2020. https://theconversation.com/how-you-can-help-not-harm-wild-animals-recovering-from-bushfires-131385.

Parsons, Edward CM, E. C. M. Parsons, A. Bauer, M. P. Simmonds, Andrew John Wright, and D. McCafferty. *An Introduction to Marine Mammal Biology and Conservation*. Burlington, MA: Jones & Bartlett Publishers, 2013

Paxton, Kristina L., Esther Sebastián-González, Justin M. Hite, Lisa H. Crampton, David Kuhn, and Patrick J. Hart. "Loss of Cultural Song Diversity and the Convergence of Songs in a Declining Hawaiian Forest Bird Community." *Royal Society Open Science* 6, no. 8 (2019): 190719.

Pedler, Reece D. "The Impacts of Abandoned Mining Shafts: Fauna Entrapment in Opal Prospecting Shafts at Coober Pedy, South Australia." *Ecological Management & Restoration* 11, no. 1 (2010): 36–42.

Perrig, Paula L., Emily D. Fountain, Sergio A. Lambertucci, and Jonathan N. Pauli. "Demography of Avian Scavengers after Pleistocene Megafaunal Extinction." *Scientific Reports* 9, no. 1 (2019): 1–9.

Philip, Justine. "Living Blanket, Water Diviner, Wild Pet: A Cultural History of the Dingo." *The Conversation*, August 6, 2017. https://theconversation.com/living-blanket-water -diviner-wild-pet-a-cultural-history-of-the-dingo-80189.

Phillips, D., and H. Nash. *Captive or Forever Free? The Condor Question."* San Francisco: Friends of The Earth, 1981.

Phippen, J. Weston. "'Kill Every Buffalo You Can! Every Buffalo Dead Is an Indian Gone'" *Atlantic*, May 13, 2016. https://www.theatlantic.com/national/archive/2016/05/the -buffalo-killers/482349/.

Pierce, J. Kingston. "Tusko the Elephant Rampages through Sedro-Woolley on May 15, 1922." HistoryLink.org, February 22, 2003. https://www.historylink.org/File/5270.

Pierotti, Raymond John, and Brandy R. Fogg. *The First Domestication: How Wolves and Humans Coevolved.* New Haven: Yale University Press, 2017.

Platts, Philip J., Suzanna C. Mason, Georgina Palmer, Jane K. Hill, Tom H. Oliver, Gary D. Powney, Richard Fox, and Chris D. Thomas. "Habitat Availability Explains Variation in Climate-driven Range Shifts Across Multiple Taxonomic groups." *Scientific Reports* 9, no. 1 (2019): 1–10.

Plummer, Kate E., Gavin M. Siriwardena, Greg J. Conway, Kate Risely, and Mike P. Toms. "Is Supplementary Feeding in Gardens a Driver of Evolutionary Change in a Migratory Bird Species?" *Global Change Biology* 21, no. 12 (2015): 4353–4363.

Plumwood, Val. "Human Vulnerability and the Experience of Being Prey." *Quadrant* 39, no. 3 (1995): 29.

Plumwood, Val. "Integrating Ethical Frameworks for Animals, Humans, and Nature: A Critical Feminist Eco-socialist Analysis." *Ethics & the Environment* 5, no. 2 (2000): 285–322.

Plumwood, Val. "Tasteless: Towards a Food-based Approach to Death." *Environmental Values* 17, no. 3 (2008): 323–330.

Pobiner, Briana. "Meat-Eating Among the Earliest Humans." *American Scientist* 104, no. 2 (2016): 110–117.

Predator Free NZ. "Backyard Trapping." Accessed on November 13, 2020. https://predatorfreenz.org/get-involved/backyard-trapping/.

Preston, Stephanie D., and Frans B. M. de Waal. "Empathy: Its Ultimate and Proximate Bases." *Behavioral and Brain Sciences* 25, no. 1 (2002).

Prokosch, Jorinde, Zephne Bernitz, Herman Bernitz, Birgit Erni, and Res Altwegg. "Are Animals Shrinking Due to Climate Change? Temperature-mediated Selection on Body Mass in Mountain Wagtails." *Oecologia* 189, no. 3 (2019): 841–849.

Public Law 88-577 (16 U.S.C. 1131-1136) 88th Congress, Second Session September 3, 1964. "To establish a National Wilderness Preservation System for the permanent good of the whole people, and for other purposes."

Radford, James Q., John CZ Woinarski, Sarah Legge, Marcus Baseler, Joss Bentley, Andrew A. Burbidge, Michael Bode, et al. "Degrees of Population-level Susceptibility of Australian Terrestrial Non-volant Mammal Species to Predation by the Introduced Red Fox (*Vulpes vulpes*) and Feral Cat (*Felis catus*)." *Wildlife Research* 45, no. 7 (2018): 645–657.

Randi, Ettore, Vittorio Lucchini, Mads Fjeldsø Christensen, Nadia Mucci, Stephan M. Funk, Gaudenz Dolf, and Volker Loeschcke. "Mitochondrial DNA Variability in Italian and East European Wolves: Detecting the Consequences of Small Population Size and Hybridization." *Conservation Biology* 14, no. 2 (2000): 464–473.

Reddiex, Ben, David M. Forsyth, Eve McDonald-Madden, Luke D. Einoder, Peter A. Griffioen, Ryan R. Chick, and Alan J. Robley. "Control of Pest Mammals for Biodiversity Protection in Australia. I. Patterns of Control and Monitoring." *Wildlife Research* 33, no. 8 (2007): 691–709.

Reed-Guy, Sarah, Connor Gehris, Meng Shi, and Daniel T. Blumstein. "Sensitive Plant (*Mimosa pudica*) Hiding Time depends on Individual and State." *PeerJ* 5 (2017): e3598.

Regan, Tom. *The Case for Animal Rights*. Berkeley and Los Angeles: University of California Press, 2004

Regehr, Eric V., Kristin L. Laidre, H. Resit Akçakaya, Steven C. Amstrup, Todd C. Atwood, Nicholas J. Lunn, Martyn Obbard, Harry Stern, Gregory W. Thiemann, and Øystein Wiig. "Conservation Status of Polar Bears (*Ursus maritimus*) in Relation to Projected Sea-ice Declines." *Biology Letters* 12, no. 12 (2016): 20160556.

Riggio, Jason, Jonathan E. M. Baillie, Steven Brumby, Erle Ellis, Christina M. Kennedy, James R. Oakleaf, Alex Tait, et al. "Global Human Influence Maps Reveal Clear Opportunities in Conserving Earth's Remaining Intact Terrestrial Ecosystems." *Global Change Biology* (2020): 4344–4356.

Rinker, David C., Natalya K. Specian, Shu Zhao, and John G. Gibbons. "Polar Bear Evolution Is Marked by Rapid Changes in Gene Copy Number in Response to Dietary Shift." *Proceedings of the National Academy of Sciences* 116, no. 27 (2019): 13446–13451.

Robb, Gillian N., Robbie A. McDonald, Dan E. Chamberlain, S. James Reynolds, Timothy JE Harrison, and Stuart Bearhop. "Winter Feeding of Birds Increases Productivity in the Subsequent Breeding Season." *Biology Letters* 4, no. 2 (2008): 220–223.

Roca, Irene T., Louis Desrochers, Matteo Giacomazzo, Andrea Bertolo, Patricia Bolduc, Raphaël Deschesnes, Charles A. Martin, Vincent Rainville, Guillaume Rheault, and Raphaël Proulx. "Shifting Song Frequencies in Response to Anthropogenic Noise: A Meta-analysis on Birds and Anurans." *Behavioral Ecology* 27, no. 5 (2016): 1269–1274.

Rohwer, Yasha. "A Duty to Cognitively Enhance Animals." *Environmental Values* 27, no. 2 (2018): 137–158.

Rohwer, Yasha. "Gene Drives, Species, and Compassion for Individuals in Conservation Biology." *Ethics, Policy & Environment*, November 18, 2020. https://www.tandfonline.com/doi/full/10.1080/21550085.2020.1848184.

Rohwer, Yasha, and Emma Marris. "Is There a Prima Facie Duty to Preserve Genetic Integrity in Conservation Biology?" *Ethics, Policy & Environment* 18, no. 3 (2015): 233–247.

Rohwer, Yasha, and Emma Marris. "Renaming Restoration: Conceptualizing and Justifying the Activity as a Restoration of Lost Moral Value Rather Than a Return to a Previous State." *Restoration Ecology* 24, no. 5 (2016): 674–679.

Rohwer, Yasha, and Emma Marris. "An Analysis of Potential Ethical Justifications for Mammoth De-extinction and a Call for Empirical Research." *Ethics, Policy & Environment* 21, no. 1 (2018): 127–142.

Rohwer, Yasha, and Emma Marris. "Clarifying Compassionate Conservation with Hypotheticals: Response to Wallach et al. 2018." *Conservation Biology* 33, no. 4 (2019): 781–783.

Rolston, Holmes. *Environmental Ethics: Duties to and Values in the Natural World: Book Summary.* Philadelphia: Temple University Press, 1989.

Rothfels, Nigel. *Savages and Beasts: The Birth of the Modern Zoo.* Baltimore: The Johns Hopkins University Press, 2002.

Routley, Richard. "Is There a Need for a New, an Environmental Ethic?" *In Proceedings of the XVth World Congress of Philosophy* 1 (1973): 205–210.

Royal New Zealand Society for the Prevention of Cruelty to Animals Incorporated. "1080—What Is It, and What Can Be Done about It?" January 7, 2019. https://www.spca.nz /news-and-events/news-article/1080-what-is-it-and-what-can-be-done-about-it.

Royal New Zealand Society for the Prevention of Cruelty to Animals Incorporated. "SPCA Prosecutes Woman for Neglecting Her Pet Rabbit" February 12, 2020. https://www.spca .nz/news-and-events/news-article/neglectedrabbit.

San Diego Zoo Global Library. "California Condor (*Gymnogyps californianus*) Fact Sheet." Accessed November 11, 2020. https://ielc.libguides.com/sdzg/factsheets/californiacondor.

Sandler, Ronald L. *The Ethics of Species: An Introduction.* United Kingdom: Cambridge University Press, 2012.

Sandler, Ronald. *Environmental Ethics: Theory in Practice.* New York: Oxford University Press, 2017.

Sandler, Ronald. "The Ethics of Genetic Engineering and Gene Drives in Conservation." *Conservation Biology* 34, no. 2 (2020): 378–385.

Sandom, Christopher, Søren Faurby, Brody Sandel, and Jens-Christian Svenning. "Global Late Quaternary Megafauna Extinctions Linked to Humans, Not Climate Change." *Proceedings of the Royal Society B: Biological Sciences* 281, no. 1787 (2014): 20133254.

Santa Monica Mtns. Twitter, November 17, 2018. https://twitter.com/SantaMonicaMtns/status /1063850186709975040?s=20.

Santos-Fita, Dídac, Eduardo J. Naranjo, Erin IJ Estrada, Ramón Mariaca, and Eduardo Bello. "Symbolism and Ritual Practices Related to Hunting in Maya Communities from Central Quintana Roo, Mexico." *Journal of Ethnobiology and Ethnomedicine* 11, no. 1 (2015): 71.

Sapolsky, Robert M., and Lisa J. Share. "A Pacific Culture among Wild Baboons: Its Emergence and Transmission." *PLoS Biol* 2, no. 4 (2004): e106.

Schilthuizen, Menno, Lúcia P. Santos Pimenta, Youri Lammers, Peter J. Steenbergen, Marco Flohil, Nils G. P. Beveridge, Pieter T. van Duijn, et al. "Incorporation of an Invasive Plant into a Native Insect Herbivore Food Web." *PeerJ* 4 (2016): e1954.

Schwartz, Dominique. "Death Row Dingoes Set to Be the Environmental Saviour of Great Barrier Reef's Pelorus Island." *ABC News*, July 22, 2016. https://www.abc.net.au/news /2016-07-23/dingoes-set-to-become-pelorus-island-environmental-saviour/7652424.

Scott, J. Michael, John A. Wiens, Beatrice Van Horne, and Dale D. Goble. *Shepherding Nature: The Challenge of Conservation Reliance*. New York: Cambridge University Press, 2020.

Seattle Times. "Conservation Gift to Honor Hansa." June 11, 2007. https://www.seattletimes .com/seattle-news/conservation-gift-to-honor-hansa/.

SeaWorld. "Orca Encounter." https://seaworld.com/orlando/shows/orca-encounter/. Accessed November 10, 2020.

SeaWorld Entertainment. "2019 Annual Report." Accessed November 11, 2020. https://s1 .q4cdn.com/392447382/files/doc_financials/Annual%20Reports/Annual/SEAS-2019 -Annual-Report.pdf.

Shepard Jr, Glenn Harvey. "Hunting in Amazonia." In *Encyclopaedia of the History of Science, Technology, and Medicine in Non-Western Cultures*, edited by Helaine Selin. New York: Springer, Dordrecht, 2014.

Shepard Jr, Glenn H., Klaus Rummenhoeller, Julia Ohl-Schacherer, and Douglas W. Yu. "Trouble in Paradise: Indigenous Populations, Anthropological Policies, and Biodiversity Conservation in Manu National Park, Peru." *Journal of Sustainable Forestry* 29, no. 2-4 (2010): 252–301.

Shipek, Florence C. "A Native American Adaptation to Drought: The Kumeyaay as Seen in the San Diego Mission Records 1770–1798." *Ethnohistory* 28, no. 4 (1981): 295–312.

Shumaker, Robert W., Kristina R. Walkup, and Benjamin B. Beck. *Animal Tool Behavior: The Use and Manufacture of Tools by Animals*. Baltimore: The Johns Hopkins University Press, 2011.

Siebert, Charles. "Zoos Called It a 'Rescue.' But Are the Elephants Really Better Off?" *New York Times*, July 9, 2019. https://www.nytimes.com/2019/07/09/magazine/elephants -zoos-swazi-17.html.

Singer, Michael C., and Camille Parmesan. "Lethal Trap Created by Adaptive Evolutionary Response to an Exotic Resource." *Nature* 557, no. 7704 (2018): 238–241.

Singer, Peter. *Animal Liberation: Towards an End to Man's Inhumanity to Animals.* United Kingdom: Granada Publishing Ltd., 1977.

Skoglund, Pontus, Erik Ersmark, Eleftheria Palkopoulou, and Love Dalén. "Ancient Wolf Genome Reveals an Early Divergence of Domestic Dog Ancestors and Admixture into High-Latitude Breeds." *Current Biology* 25 (2015): 1515–1519.

Slinker, Bryan and Rob Liddell. "Op-Ed: Zoos Play a Vital Role Protecting Wild Elephants and Their Habitat." *Seattle Times.* December 8, 2012. https://www.seattletimes.com /opinion/op-ed-zoos-play-a-vital-role-protecting-wild-elephants-and-their-habitat/.

Smart, Amy. "Grizzly Bears Move North in High Arctic as Climate Change Expands Range." *National Post*, December 14, 2019. https://nationalpost.com/news/canada/grizzly -bears-move-north-in-high-arctic-as-climate-change-expands-range.

Smith, Anna. "An Indigenous Effort to Return Condors to the Pacific Northwest Nears Its Goal." *Audubon*, November 5, 2020. https://www.audubon.org/news/an-indigenous -effort-return-condors-pacific-northwest-nears-its-goal.

Smith, Blanton. "Possum Pics Disgust." *Taranaki Daily News*, August 2, 2012. https://www .stuff.co.nz/taranaki-daily-news/news/7392131/Possum-pics-disgust.

Smith, Bradley. *The Dingo Debate: Origins, Behaviour and Conservation.* Australia: CSIRO Publishing, 2015.

Smith, Felisa A., Catalina P. Tomé, Emma A. Elliott Smith, S. Kathleen Lyons, Seth D. Newsome, and Thomas W. Stafford. "Unraveling the Consequences of the Terminal Pleistocene Megafauna Extinction on Mammal Community Assembly." *Ecography* 39, no. 2 (2016): 223–239.

Smith, Joe. "Winter Bird Feeding: Good or Bad for Birds?" Cool Green Science, January 5, 2015. https://blog.nature.org/science/2015/01/05/winter-bird-feeding-good-or-bad-for -birds/.

Snyder, Noel F. R., and Helen A. Snyder. "Biology and Conservation of the California Condor." In *Current Ornithology*, vol 6, edited by D. M. Power, 175–267. Boston, MA: Springer, 1989.

Soulé, Michael E. "What Is Conservation Biology?" *BioScience* 35, no. 11 (1985): 727–734.

Sports Illustrated. "Last Chance for the Condor." March 23, 1987. https://vault.si.com/vault /1987/03/23/last-chance-for-the-condor-with-only-one-of-the-species-left-in-the-wild -a-remarkable-bird-struggles-for-survival.

Stallcup, R. "Farewell, Skymaster." *Point Reyes Bird Observatory Newsletter* 53 (1981): 10.

Stanford Encyclopedia of Philosophy, Winter 2017 edition. "Moral Particularism," by Jonathan Dancy. Last modified September 22, 2017, https://plato.stanford.edu/archives /win2017/entries/moral-particularism.

Stanford Encyclopedia of Philosophy, Spring 2018 edition. "Value Pluralism," by Elinor Mason. Last modified February 7, 2018. https://plato.stanford.edu/archives/spr2018 /entries/value-pluralism/.

Stanford Encyclopedia of Philosophy, Fall 2020 edition. "Hobbes's Moral and Political Philosophy," by Sharon A. Lloyd and Susanne Sreedhar. Last modified April 30, 2018. https://plato.stanford.edu/archives/fall2020/entries/hobbes-moral/.

State of New York Court of Appeals, *Nonhuman Rights Project, Inc., on Behalf of Tommy v. Patrick C. Lavery, &c., et al/Nonhuman Rights Project, Inc., on Behalf of Kiko v. Carmen Presti et al*, Motion No. 2018-268.

Steinfurth, Antje. "Marine Ecology and Conservation of the Galápagos Penguin, *Spheniscus mendiculus.*" PhD diss., Christian-Albrecht University of Kiel, 2007. https://d-nb. info/1019952164/34.

Suárez-Rodríguez, Monserrat, Isabel López-Rull, and Constantino Macias Garcia. "Incorporation of Cigarette Butts into Nests Reduces Nest Ectoparasite Load in Urban Birds: New Ingredients for an Old Recipe?" *Biology Letters* 9, no. 1 (2013): 20120931.

Svartberg, Kenth, Ingrid Tapper, Hans Temrin, Tommy Radesäter, and Staffan Thorman. "Consistency of Personality Traits in Dogs." *Animal Behaviour* 69, no. 2 (2005): 283–291.

Sydney Morning Herald. "Val Plumwood Died of Natural Causes: Friend." March 6, 2008. smh.com.au/environment/val-plumwood-died-of-natural-causes-friend-20080306 -gds455.html.

Taylor, H. J. "The Last Survivor." *University of California Chronicle*, January, 1931, Vol. XXXIII, No. I and January, 1932, Vol. XXXIV, No. 1. Accessed at https://www.yosemite.ca .us/library/the_last_survivor/.

Terrill, Ceiridwen. *Part Wild: One Woman's Journey with a Creature Caught Between the Worlds of Wolves and Dogs.* New York: Simon and Schuster, 2011.

The Condor Cave. "The Loss of AC-9." Facebook. July 11, 2017. https://www.facebook .com/notes/the-condor-cave/the-loss-of-ac-9/1734698233211159/.

The Whale Sanctuary Project. "We're building a model sanctuary where captive whales and dolphins can be rehabilitated and live permanently in their natural environment." Home page. Accessed November 11, 2020. https://whalesanctuaryproject.org/.

Thomas, Chris D. *Inheritors of the Earth: How Nature Is Thriving in an Age of Extinction.* United Kingdom: Hachette, 2017.

Thomson, Judith Jarvis. "The Trolley Problem." *Yale Law Journal* 94, no. 6 (1985): 1395–415. Accessed December 2, 2020. doi:10.2307/796133.

Urban, Mark C. "Climate-tracking Species Are Not Invasive." *Nature Climate Change* 10 (2020): 382–384.

U.S. Fish & Wildlife Service. "California Condor Population Information." February 18, 2020. https://www.fws.gov/cno/es/CalCondor/Condor-population.html.

U.S. Fish & Wildlife Service. "Barred Owl Study Update." Accessed December 2, 2020. https://www.fws.gov/oregonfwo/articles.cfm?id=149489616.

Van Dine, Delos Lewis. *Mosquitoes in Hawaii.* Hawaiian Gazette Company, 1904.

Vernes, Sonja C., and Gerald S. Wilkinson. "Behaviour, Biology and Evolution of Vocal Learning in Bats." *Philosophical Transactions of the Royal Society B* 375, no. 1789 (2020): 20190061.

Vice. "DJ Khaled Reveals 'Major Key' Album Cover Which Is the Greatest Album Cover Maybe Ever." Noisy. June 24, 2016. https://www.vice.com/en/article/rdzmdq/dj -khaled-reveals-major-key-album-cover-which-is-greatest-album-cover-maybe-ever.

Visser, Ingrid N. "Prolific Body Scars and Collapsing Dorsal Fins on Killer Whales (*Orcinus orca*) in New Zealand Waters." *Aquatic Mammals* 24 (1998): 71–82.

Wallach, Arian D., Marc Bekoff, Chelsea Batavia, Michael Paul Nelson, and Daniel Ramp. "Summoning Compassion to Address the Challenges of Conservation." *Conservation Biology* 32, no. 6 (2018): 1255–1265.

Wallach, Arian D., Erick Lundgren, Chelsea Batavia, Michael Paul Nelson, Esty Yanco, Wayne L. Linklater, Scott P. Carroll, et al. "When All Life Counts in Conservation." *Conservation Biology* 34 (2019): 997–1007.

Walsh, L. L., and P. K. Tucker. "Contemporary range expansion of the Virginia opossum (Didelphis virginiana) impacted by humans and snow cover." *Canadian Journal of Zoology* 96, no. 2 (2018): 107–115.

Walton, Alexander, and Amy L. Toth. "Variation in Individual Worker Honey Bee Behavior Shows Hallmarks of Personality." *Behavioral Ecology and Sociobiology* 70, no. 7 (2016): 999–1010.

Waters, Hannah. "Knights in Shining Fur: The Fight to Save Australia's Littlest Penguins." *Audubon*, September 25, 2015. https://www.audubon.org/news/knights-shining-fur -fight-save-australias-littlest-penguins.

Weaver, Caity. "Justin Bieber Would Like to Reintroduce Himself." *GQ*, February 11, 2016. https://www.gq.com/story/justin-bieber-gq-interview?

Weiskopf, Sarah R., Madeleine A. Rubenstein, Lisa G. Crozier, Sarah Gaichas, Roger Griffis, Jessica E. Halofsky, Kimberly J. W. Hyde, et al. "Climate Change Effects on Biodiversity, Ecosystems, Ecosystem Services, and Natural Resource Management in the United States." *Science of the Total Environment* 773 (2020): 137782.

Wells, Herbert George. *The Island of Doctor Moreau.* New York: Bantam, 2005.

White, Lynn. "The Historical Roots of Our Ecologic Crisis." *Science* 155, no. 3767 (1967): 1203–1207.

Wiig, Ø., Amstrup, S., Atwood, T., Laidre, K., Lunn, N., Obbard, M., Regehr, E. & Thiemann, G. 2015. "*Ursus maritimus.* The IUCN Red List of Threatened Species 2015," e.T22823A14871490. https://dx.doi.org/10.2305/IUCN.UK.2015-4.RLTS.T22823A1487 1490.en. Downloaded on 11 November 2020.

Wilmshurst, Janet M., Terry L. Hunt, Carl P. Lipo, and Atholl J. Anderson. "High-Precision Radiocarbon Dating Shows Recent and Rapid Initial Human Colonization of East Polynesia." *Proceedings of the National Academy of Sciences* 108, no. 5 (2011): 1815–1820.

Winsor, Morgan. "CITES Stops Short of Outright Ban on Zoo Trade in Wild African Elephants." *ABC News.* August 28, 2019. https://abcnews.go.com/International/cites -stops-short-outright-ban-zoo-trade-wild/story?id=65239933.

Winter, Kawika B. "A Hawaiian Renaissance That Could Save the World: This Archipelago's Society Before Western Contact Developed a Large, Self-sufficient Population, Yet

Imposed a Remarkably Small Ecological Footprint." *American Scientist* 107, no. 4 (2019): 232–240.

Woodland Park Zoo. "Young Wolves Join Northern Trail." April 8, 2011. https://blog.zoo .org/2011/04/young-wolves-join-northern-trail.html.

Woodland Park Zoo. "Update: Learning More about Watoto." October 7, 2014. https:// blog.zoo.org/2014/10/update-learning-more-about-watoto.html?utm_source=feed burner&utm_medium=feed&utm_campaign=Feed:+WoodlandParkZBlog+(Wood land+Park+Zoo+Blog+%7C+Naturally+Inspiring).

Woodland Park Zoological Society, IRS form 990 for 2007. Accessed at https://projects .propublica.org/nonprofits/display_990/916070005/2009_01_EO%2F91-6070005_990 _200712.

Žampachová, Barbora, Barbora Kaftanová, Hana Šimánková, Eva Landová, and Daniel Frynta. "Consistent Individual Differences in Standard Exploration Tasks in the Black Rat (*Rattus rattus*)." *Journal of Comparative Psychology* 131, no. 2 (2017): 150.

Zealandia. *Annual Report 2018/19.* https://www.visitzealandia.com/Portals/0/Annual%20 Report%202018-2019_Web_size_smaller.pdf.

Ziegler, Alan C. *Hawaiian Natural History, Ecology, and Evolution.* Honolulu: University of Hawaii Press, 2002.

Zipple, Matthew N., Jackson H. Grady, Jacob B. Gordon, Lydia D. Chow, Elizabeth A. Archie, Jeanne Altmann, and Susan C. Alberts. "Conditional Fetal and Infant Killing by Male Baboons." *Proceedings of the Royal Society B: Biological Sciences* 284, no. 1847 (2017): 20162561.

INDEX

A NOTE ON THE AUTHOR

EMMA MARRIS is an award-winning journalist whose writing on science and the environment has appeared in the *New York Times*, the *Atlantic*, *National Geographic*, *Wired*, *Outside*, *High Country News*, and many other publications, including *The Best American Science and Nature Writing*. Her previous book, *Rambunctious Garden*, argued for a new conservation philosophy for the 21st century. Her TED Talk about redefining nature has over 1.4 million views. She is based in Klamath Falls, Oregon.